OXFORD MEDICAL PUBLICATIONS

MUSCLE DAMAGE

MUSCLE DAMAGE

Edited by

STANLEY SALMONS

Professor of Medical Cell Biology
Director, British Heart Foundation
Skeletal Muscle Assist Research Group
Department of Human Anatomy and Cell Biology
University of Liverpool
UK

OXFORD NEW YORK TOKYO
OXFORD UNIVERSITY PRESS
1997

Oxford University Press, Great Clarendon Street, Oxford OX2 6DP
Oxford   New York
Athens   Auckland   Bangkok   Bombay   Buenos Aires
Calcutta   Cape Town   Dar es Salaam   Delhi   Florence   Hong Kong
Istanbul   Karachi   Kuala Lumpur   Madras   Madrid   Melbourne
Mexico City   Nairobi   Paris   Singapore   Taipei   Tokyo   Toronto
and associated companies in
Berlin   Ibadan

Oxford is a trade mark of Oxford University Press

Published in the United States
by Oxford University Press Inc., New York

A catalogue record for this book is available from the British Library

Library of Congress Cataloging in Publication Data
(Data available)

ISBN 0 19 262753 8

Typeset by Hewer Text Composition Services, Edinburgh
Printed in Great Britain by
Biddles Ltd, Guildford & King's Lynn

To my wife Paula, whose dismay at my propensity for accepting new assignments is more than matched by her loyal support whenever such a project is under way.

Preface

In August 1992, Professor Ugo Carraro, editor of the journal *Basic and Applied Myology*, invited me to edit a special issue of his journal on the subject of muscle damage. My initial reaction was that I had enough to do. I was, however, intrigued enough to look at what it might involve. Papers would be needed on the damaging effects of exercise, stretch, and stimulation, on strain injuries and myopathic damage, on the definition and quantitation of damage, on the underlying mechanisms . . . as I wrote down each topic, I added the name of a potential contributor. This was encouraging: all the people on my list were known to me personally and were international authorities in their fields; if they were interested, the issue could indeed be a valuable one. As it turned out, they were enthusiastic about the project and, despite being as busy as myself, were prepared to offer papers on the timescale I had given them.

Although all the manuscripts had been submitted by invitation, I had no intention of bypassing the normal processes of peer review. This posed something of a problem, because the expert reviewers on whom I would normally have called were all in my list of contributors! I therefore proposed the following scheme: each paper would be reviewed by me and 'anonymously' by one of the other authors; papers with which I myself was associated would be reviewed by two of the other authors. This was agreed, and the reviews were, as I had anticipated, critical but constructive. The manuscripts were duly revised and submitted to the publishers.

The special issue came out at the beginning of 1994 (*Basic and Applied Myology*, **4**, (1), 3–112). It was well received: Ugo Carraro told me that he had received more requests for this one issue than for any preceding issue of the journal—in fact he had totally exhausted his stock of copies. *Basic and Applied Myology* was, however, a young journal, still establishing its reputation. It had a small subscription readership and it was not yet listed in the main indexing journals. There was obviously a wider audience for this material and I felt some obligation to my contributors to seek it. I had in mind a monograph on the subject, and suggested the idea to Ugo. He was very supportive, and confirmed that the publishers of the journal would be willing to grant the necessary permissions. I approached my authors and was again gratified at their enthusiastic response: all were willing to put in hand the appropriate revisions and to bring their contributions up-to-date.

The book that has emerged from this process differs in three major ways from the journal issue in which it had its origins. Firstly, the authors—good as their word—have ensured that the material that went to press was as up-to-the-minute as it could be. Secondly, additional chapters have been commissioned and included at the suggestion of the publisher's reviewers. Thirdly, with the indulgence of the authors, I have edited the material in an effort to ensure clarity, accessibility, and, as far as possible, a degree of stylistic consistency. In the process I have been able to introduce a useful degree of cross-referencing.

So what is this book about? To some individuals, muscle damage may mean nothing more than transient discomfort after unaccustomed exercise. To others, it may mean a debilitating, and possibly progressive, disease that compromises their quality of life. To yet others, it may mean enforced rest from a physically demanding occupation or from competitive sport. Thus, muscle damage has a cost, to the individual and to society, and treating or preventing it makes demands on the time and attention of practitioners of sports, occupational, and clinical medicine. This being so, it is surprising how little we know about the nature of muscle damage, how it is precipitated, how it should be assessed, and how it may be prevented or its effects ameliorated.

A part of the problem is that there are many ways of studying muscle damage, and damage tends to be defined within the context of the technique that is being used. Thus a physiologist might see damage as a loss of force-generating capacity, a clinician might look for the appearance of creatine kinase in the serum, a histologist, for invasion by mononuclear cells, and so on. This leads to difficulties in relating studies based on different techniques or experimental models, for there is often precious little correlation between either the time course or the severity of damage measured in these various ways. A further problem posed by this diversity is that the published literature tends to be distributed among a variety of discipline-based journals. There is nothing intrinsically wrong with this: indeed, it is essential if the field is to continue to benefit from the application of techniques and progress from parallel research areas. However, it does underline the need for a book such as this to bring together in one place the different insights provided by these diverse approaches.

With these considerations in mind I have avoided any attempt to eliminate overlap between chapters—in fact I have regarded any such overlap as a merit. It is quite possible to read the chapters out of sequence, but readers new to the field could do worse than to read them in the order presented, which has a certain underlying logic. What should then emerge is a timely and comprehensive statement of what we know about muscle damage and, more importantly, an indication of what we have yet to learn.

My thanks go to the authors who contributed their time and expertise to this endeavour, to my secretary, Glenis Crompton, who helped to keep my head above water, and to my colleagues Jonathan Jarvis, Adam Shortland, Hans Degens, Gus Tang, Hazel Sutherland, Stephen Gilroy, and Michelle Hastings, who ensured that the time I spent on this book was still a productive one for our laboratory.

Liverpool   S.S.
December 1996

Contents

List of contributors

P.R. Dop Bär
Professor of Experimental Neurology
Department of Neurology
University of Utrecht
University Hospital AZU
Postbox 85500
3508 GA Utrecht
The Netherlands

Thomas M. Best
Assistant Professor
Department of Family Medicine and
  Orthopaedic Surgery
University of Wisconsin Medical Center
Madison, WI 53711
USA

Anke L. Bootsma
Associate Lecturer
Department of Cell Biology
University of Utrecht
University Hospital AZU
Postbox 85500
3508 GA Utrecht
The Netherlands

Susan V. Brooks
Assistant Professor of Physiology and
  Assistant Research Scientist
University of Michigan
Institute of Gerontology
300 N. Ingalls
Ann Arbor, MI 48109–2007
USA

Adonella Bruson
Associate Fellow
Department of Biomedical Sciences
University of Padova
Via Trieste 75
I-35131 Padova
Italy

Marcello Cantini
Senior Scientist
Department of Biomedical Sciences
University of Padova
Via Trieste 75
I-35131 Padova
Italy

Ugo Carraro
Associate Professor of General
  Pathology
Department of Biomedical Sciences and
  CNR Unit for Muscle Biology and
  Physiopathology
University of Padova
Via Trieste 75
I-35131 Padova
Italy

Claudia Catani
Senior Scientist
Department of Biomedical Sciences and
  CNR Unit for Muscle Biology and
  Physiopathology
University of Padova
Via Trieste 75
I-35131 Padova
Italy

Valerie M. Cox
Lecturer in Exercise Physiology
Department of Biological Sciences
Science Park
University of Durham
Durham DH1 3LE
UK

David Y. Downham
Senior Lecturer
Department of Statistics and
  Computational Mathematics
University of Liverpool
Liverpool L69 3BX
UK

John A. Faulkner
Professor of Physiology
Associate Director and Research
  Scientist
University of Michigan
Institute of Gerontology
300 N. Ingalls
Ann Arbor, MI 48109–2007
USA

Jan Fridén
Associate Professor of Hand Surgery
Department of Hand Surgery
Sahlgrenska University Hospital
University of Göteborg
S-413 45 Göteborg
SWEDEN

William E. Garrett, Jr
Professor
Division of Orthopaedic Surgery
Duke University Medical Center
Durham, NC
USA

Carl T. Hasselman
Resident
Department of Orthopaedic Surgery
University of Pittsburgh Medical Center
Pittsburgh, PA 15213
USA

Timothy R. Helliwell
Senior Lecturer in Pathology and
  Honorary Consultant Pathologist
Department of Pathology
The Royal Liverpool University
  Hospitals
Liverpool L69 3BX
UK

Malcom J. Jackson
Professor of Cellular Pathology
Muscle Research Centre
Department of Medicine
University of Liverpool
Liverpool L69 3BX
UK

Sandra C.J.M. Jacobs
Research Fellow
Departments of Neurology and Cell
  Biology
University of Utrecht
University Hospital AZU
Postbox 85500
3508 GA Utrecht
The Netherlands

Jonathan C. Jarvis
Postdoctoral Research Fellow
Associate Director, British Heart
  Foundation Skeletal Muscle Assist
  Research Group
Department of Human Anatomy and
  Cell Biology
University of Liverpool
Liverpool L69 3BX
UK

David A. Jones
Professor and Head of School
School of Sport and Exercise Sciences
The University of Birmingham
Edgbaston
Birmingham B15 2TT
UK

Jan Lexell
Associate Professor of Experimental
  Neurology
Director, Neuromuscular Research
  Laboratory
Department of Rehabilitation
Lund University Hospital
Orupssjukhuset
S-243 85 Höör
Sweden

Richard L. Lieber
Professor of Orthopaedics and
  Bioengineering
Departments of Orthopaedics and
  Bioengineering
Biomedical Sciences Graduate Group
University of California San Diego
  School of Medicine and Veterans
  Administration Medical Center
U.C. 3350, La Jolla Village Drive
San Diego, CA 92161–9151
USA

Francesco Mazzoleni
Professor of Plastic Surgery
Institute of Plastic and Reconstructive
  Surgery
University of Padova
Via Trieste 75
I-35131 Padova
Italy

Anne McArdle
Postdoctoral Research Fellow
Muscle Research Centre
Department of Medicine
University of Liverpool
Liverpool L69 3BX
UK

Jaap C. Reijneveld
Neurologist in training
Department of Neurology
University of Utrecht
University Hospital AZU
Postbox 85500
3508 GA Utrecht
The Netherlands

Corrado Rizzi
Plastic Surgeon
Institute of Plastic and Reconstructive
  Surgery and
Department of Biomedical Sciences
University of Padova
Via Trieste 75
I-35131 Padova
Italy

Marco Rossi
Pædiatric Surgeon
Department of Biomedical Sciences
University of Padova
Via Trieste 75
I-35131 Padova
Italy

Katia Rossini
Associate Fellow
Department of Biomedical Sciences
University of Padova
Via Trieste 75
I-35131 Padova
Italy

Joan M. Round
Research Fellow
School of Sport and Exercise Sciences
The University of Birmingham
Edgbaston
Birmingham B15 2TT
UK

Stanley Salmons
Professor of Medical Cell Biology
Director, British Heart Foundation
  Skeletal Muscle Assist Research
  Group
Department of Human Anatomy and
  Cell Biology
University of Liverpool
Liverpool L69 3BX
UK

Marco Sandri
Associate Fellow
Department of Biomedical Sciences
University of Padova
Via Trieste 75
I-35131 Padova
Italy

Hazel Sutherland
Research Assistant
British Heart Foundation Skeletal
    Muscle Assist Research Group
Department of Human Anatomy and
    Cell Biology
University of Liverpool
Liverpool L69 3BX
UK

J.H. John Wokke
Professor of Neuromuscular Diseases
Department of Neurology
University of Utrecht
University Hospital AZU
Postbox 85500
3508 GA Utrecht
The Netherlands

List of abbreviations

A23187	a calcium ionophore
ATP	adenosine 5'-triphosphate
bFGF	basic fibroblast growth factor
CA III	carbonic anhydrase III
CK	creatine kinase
CK-MM	muscle-specific isoenzyme of creatine kinase
CNS	central nervous system
CT	computerized tomography
^{51}Cr-EDTA	^{51}Cr-labelled ethylene diamine tetra-acetic acid
^{51}Cr-RBC	^{51}Cr-labelled red blood cells
DNP	2,4 dinitrophenol
DOMS	delayed-onset muscle soreness
ECF	extracellular fluid
EDL	extensor digitorum longus
EMG	electromyogram
ESR	electron spin resonance
H & E	haematoxylin and eosin
HMGRI	3–hydroxy-3–methylglutaryl coenzyme A reductase inhibitor
IBM	inclusion body myositis
ICAM-1	intercellular adhesion molecule-1
IGF-I	insulin-like growth factor-I
LDH	lactate dehydrogenase
Mb	myoglobin
MHC	myosin heavy chain
MR	magnetic resonance
MRI	magnetic resonance imaging

MTJ	muscle-tendon junction (or myotendinous junction)
NADH	nicotinamide-adenine dinucleotide
NADPH	nicotinamide-adenine dinucleotide phosphate
NCAM	neural cell adhesion molecule
NSAID	non-steroidal anti-inflammatory drug
^{31}P-MRS	phosphorus-magnetic resonance spectroscopy
PCr	phosphocreatine
PEG-SOD	polyethylene glycol superoxide dismutase
PET	proton emission tomography
P_i	inorganic phosphate
SCARMD	severe childhood autosomal recessive muscular dystrophy
SD	standard deviation
SDS PAGE	sodium dodecyl sulphate polyacrylamide gel electrophoresis
SEM	standard error of the mean
SR	sarcoplasmic reticulum
TA	tibialis anterior
TGF-β	transforming growth factor-beta
^{99}Tc^m-pyp	^{99}technetium pyrophosphate

Muscle damage induced by exercise: nature, prevention and repair

P.R. Dop Bär, Jaap C. Reijneveld, J.H. John Wokke,
Sandra C.J.M. Jacobs, and Anke L. Bootsma

1.1 Introduction

This chapter focuses on muscle damage occurring after any form of unaccustomed or high-intensity exercise. Such exercise-induced muscle damage is a very common, physiological phenomenon that can occur in normal, daily life as well as in high-intensity training. Muscle damage can also occur as a side-effect of drugs, and this type of pathological response will be considered later.

Most people are familiar with the consequences of exercise or unaccustomed motion, which include muscle pain or stiffness. Although pain is the most notorious symptom, it is not a good indicator of the amount of muscle damage: it is an indirect outcome which usually develops at a late stage (hence the name 'delayed onset muscle soreness'). It is hard to measure in an objective way, and difficult to explain or to correlate with other observations. At the other end of the spectrum of measures used to describe muscle damage lies the 'gold standard'—histological verification. Strictly speaking, this is the only method for which one can truly justify the term 'damage', as the light or electron microscope allows the observer to visualize damage to the structure of skeletal muscle, and to express it in quantitative terms.

Between the easy but subjective assessment of muscle damage on a pain scale, and the objective (but often less practicable) histological approach, there are various other ways of assessing and quantifying muscle damage. Both functional parameters (range of motion of joints, ability to exert force) and biochemical parameters (presence in the serum of enzymes or proteins that are more-or-less specific for skeletal muscle) are commonly used to assess muscle damage, often in the absence of an established quantitative relationship between that parameter and the gold standard—histological verification.

In practice, assessment of the total amount of damage in a muscle can be very difficult, if not impossible, because:

(1) different muscles and fibre types respond differently to exercise (Fridén and Lieber 1992);
(2) different muscles may contain different concentrations of enzyme or protein markers (Amelink *et al.* 1988);
(3) damage is not distributed homogeneously in a given muscle; after running, for example, more damage is observed in the proximal part of the soleus muscle of rats than in the distal part (Ogilvie *et al.* 1988; Jacobs *et al.* 1993).

Moreover, the indicator of damage used may not have a known relationship to actual loss of function or the potential for recovery.

We will use the following terminology in this chapter. We consider *pain*, *damage*, and *dysfunction* as outcomes of exercise (see also Fig. 1.1). In human studies *pain* can be assessed with a scoring system, *dysfunction* can be measured directly in terms of functional deficit, and *damage* can be measured directly by examining muscle biopsies histologically or indirectly by analysing for muscle proteins in the circulation. In animal experiments pain cannot be assessed and dysfunction may be hard to measure, so that in general only the direct and indirect measures of damage remain.

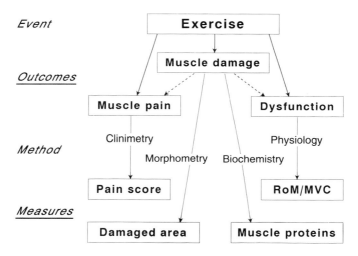

Fig 1.1 Schematic representation of the interrelationship of outcomes of exercise. The outcomes (pain, dysfunction, and damage) can be described in objective, quantitative measures that can be obtained in several ways. Note that not all arrows indicate a proven relationship: it remains to be seen whether pain and dysfunction are caused by muscle damage (dotted arrows) or via some other route. Abbreviations: RoM, range of motion of joints; MVC, maximal voluntary contraction force.

Creatine kinase (CK) activity in serum is frequently used as a marker for muscle damage. Strictly speaking, CK activity is not a measure but only an indicator that the integrity of the membrane of skeletal (or heart) muscle cells has been affected. The exact relationships between the extent of muscle damage and CK activity, and the inter-individual variability in CK response, are not clear (Clarkson *et al.* 1992). Although serum CK activity has been used to measure muscle damage (van der Meulen 1991; Volfinger *et al.* 1994), we and other authors doubt its value as an adequate quantitative marker (Nosaka and Clarkson 1992; Kuipers 1994; Komulainen *et al.* 1995). Equations for calculating the volume of damaged muscle, such as those used in cardiology to estimate infarct size, cannot be expected to be useful for skeletal muscle because of the dissimilar properties of the two muscle types. More direct methods of visualizing damage may be accessible in the future, and these may include the distribution of a labelled isotope in tissue (see Chapter 8), PET scanning, or magnetic resonance (MR) imaging (see below).

1.2   Symptoms of muscle damage

1.2.1   Pain

Prolonged or strenuous exercise can lead to several types of muscle pain. During the final stages of high-intensity work, or soon afterwards, muscle pain may be experienced, accompanied in many cases by aches, cramps, motor impairment, and fatigue, all of which gradually subside after exercise. Depletion of glycogen has been documented under these conditions and it has been assumed that the symptoms are caused by metabolic exhaustion (Sjöstrom and Fridén 1984). This interpretation is supported by similarities to the transient muscle pains, aches, and cramps induced by exercise in patients with metabolic myopathies such as myophosphorylase deficiency (McArdle's disease). Ischaemia is also regarded as a possible causative factor (Sjöstrom and Fridén 1984). Delayed-onset muscle soreness (DOMS), on the other hand, is not related to fatigue and is described as a dull, aching pain combined with tenderness and stiffness (Armstrong 1984). This type of pain is experienced in response to motion or palpation and develops gradually during the first 24 to 48 h after exercise, reaches a peak between 24 and 72 h and then subsides; usually it is no longer felt after 5–7 days (Newham *et al.* 1983; Armstrong 1984; Bobbert *et al.* 1986). Eccentric exercise (in which the muscle contracts while lengthening) is especially effective in provoking this type of muscle pain (Armstrong 1984; Miles and Clarkson 1994). The tenderness is frequently localized in the region of the distal myotendinous junction, but it can also be generalized throughout the muscle (Armstrong 1984). Despite recent studies of DOMS, its cause and its characteristic delay in onset remain unexplained. In a recent review it was postulated that an inflammatory response leads to macrophage accumulation and prostaglandin (PGE_2) release after 24 h, reaching a maximum at 48 h and gradually subsiding after 72 h (Smith 1991). PGE_2 would sensitize type III and IV pain afferents, making the muscle tender to motion and palpation. However, the use of anti-inflammatory drugs in decreasing DOMS is of questionable value (Headley *et al.* 1985; Kuipers *et al.* 1985; Donnely *et al.* 1990; Hasson *et al.* 1993). Further research will be needed to test this interesting hypothesis.

1.2.2 Dysfunction

Maximal muscle power can be reduced by 50 per cent or more after damaging exercise. Muscle strength reaches its lowest value immediately after eccentric exercise and gradually recovers over 10 days (Clarkson *et al.* 1992). This seems to be the consequence of a reduction in voluntary effort due to pain, together with a diminished capacity for generating intrinsic force, possibly as a result of damage to the sarcoplasmic reticulum (Edwards *et al.* 1977; Newham *et al.* 1983; Byrd 1992). A reduction in the ability of muscle to generate force is reflected in an increase in electrical activity of the muscle during voluntary movement performed after eccentric exercise (Newham *et al.* 1983; Berry *et al.* 1990).

The range of motion can be impaired immediately after exercise. It has been reported that the ability to flex the forearm fully is lost immediately after eccentric exercise, and that this effect is still apparent after 10 days (Clarkson *et al.* 1992; Rodenburg *et al.* 1993). Both the decrease in muscle strength and the inability to shorten muscles could be the result of over-stretching of sarcomeres by the lengthening contractions. If sarcomeres are stretched beyond the point at which the actin and myosin filaments overlap, there is a reduction in the overall number of cross-bridges that can be formed. There is some evidence to support this theory since greater losses of strength are produced by exercise at longer muscle lengths (Newham 1988; also see §2.4.1).

Muscle shortening has been observed after eccentric exercise (Clarkson and Tremblay 1988; Clarkson *et al.* 1992; Rodenburg *et al.* 1993), and is evident in the more acute resting angle of the elbow immediately after eccentric exercise involving the biceps brachii muscle. This phenomenon cannot be attributed to swelling or oedema since this is not observed earlier than 24 h after exercise (Bobbert *et al.* 1986; Clarkson *et al.* 1992). The increase in T1 and T2 relaxation time on magnetic resonance images after eccentric exercise, which is thought to reflect an increased water content of the muscle (Fleckenstein *et al.* 1988; McCully *et al.* 1988; Rodenburg *et al.* 1992*a*), is also a late phenomenon. Shortening cannot result from increased activity at rest, since there is no evidence for an increase in the resting electromyogram (Howell *et al.* 1985; Bobbert *et al.* 1986). Jones and co-workers (1987) confirmed these observations and suggested that connective tissue is involved in muscle shortening. Alternatively, abnormally high Ca^{2+} levels in the sarcoplasm, resulting from a decrease in Ca^{2+} uptake by the sarcoplasmic reticulum (Byrd 1992), may be responsible. (For further discussion of this topic, see §6.3.1.)

1.3   Other features of muscle damage

Exercise can produce an increase in the activity of several proteins in the serum (Noakes 1987), and this is generally thought to reflect an increased release or leakage from muscle. The most important enzymes, in terms of their muscle specificity, are CK and carbonic anhydrase III (CA III). A protein that can be used as a sensitive indicator of muscle damage is myoglobin (Mb).

1.3.1   Creatine kinase

The properties of this enzyme and its isoenzymes, and its use as a marker of muscle damage, have been reviewed extensively (Hortobágyi and Denahan 1989; Bär *et al.* 1990; Bär and Amelink 1992). Serum CK activity generally peaks 24–48 h after prolonged exercise (Noakes 1987). The type of exercise determines the profile of the post-exercise CK profile: after prolonged running (40–100 km) (Noakes 1987) and prolonged cross-country skiing (Takala *et al.* 1989), CK is elevated immediately, can reach peak values up to thousands of units per litre after 24–48 h and can remain elevated for up to 6–7 days. Unaccustomed eccentric exercise, even of relatively short duration, can lead to large, delayed peaks in CK activity 2 to 5 days later (Noakes 1987; Ebbeling and Clarkson 1989; Clarkson *et al.* 1992; Komulainen *et al.* 1994). Techniques that specifically demonstrate isoenzymes show that the major part of the CK rise in animals is caused by the muscle-specific isoenzyme, CK-MM (Amelink *et al.* 1988).

1.3.2   Carbonic anhydrase III and myoglobin

CA III and Mb are more specific to muscle than CK. CA III is present only in type 1 fibres (Väänänen *et al.* 1986) and Mb is present only in heart and skeletal muscle (Kagen *et al.* 1980). Assaying Mb has become easier now that a non-radioactive, nephelometric assay is available, but CA III has never achieved routine use. The profile of Mb levels and CA III activity after exercise is quite different to that of CK: CA III peaks immediately after long-distance running, whereas under identical conditions CK peaks only after 24 h (Osterman *et al.* 1985; Väänänen *et al.* 1986). The concentration of Mb shows a similar pattern to that of CA III: it peaks immediately after exercise (Nørregaard-Hansen *et al.* 1982; Driessen-Kletter *et al.* 1990; Rodenburg *et al.* 1993), making it more suitable as a marker of acute events than of chronic disease (Borleffs *et al.* 1987). After long-distance running, Mb levels increase up to 10– or 20–fold, and much higher values have been reported in cases of clinically overt rhabdomyolysis with myoglobinuria and after eccentric exercise (Nosaka and Clarkson 1992; Rodenburg *et al.* 1992*b*). In these cases, Mb remains elevated over a longer period of time.

1.3.3 Sex differences

The difference in CK release between the sexes is remarkable: women have a lower CK activity than men at rest (Meltzer 1971) and show less CK efflux after bicycle exercise (Shumate *et al.* 1979) or long-distance running (Berg and Keul 1981; Rogers *et al.* 1985). It has been suggested that oestradiol might determine these sex-linked differences (Shumate *et al.* 1979). We tested this hypothesis in an animal model in which cannulated rats exercised for 2 h on a treadmill. This exercise resulted in higher post-exercise CK values in males than in females. Ovariectomy of female rats resulted in an increase in the post-exercise CK response (Amelink and Bär 1986) and treatment of male rats with oestradiol resulted in a marked attenuation of the CK response (Bär *et al.* 1988), paralleled by a decrease in the amount of histological damage (Reijneveld *et al.* 1994). Since the muscle content of CK is the same in male and female rats (Amelink *et al.* 1988) and oestrogen treatment does not change the CK content (Dempsey *et al.* 1975; Ferrington *et al.* 1992), it was concluded that oestrogens exert a protective effect on the muscle membrane. These observations were confirmed in an *in-vitro* model based on isolated soleus muscles (Amelink *et al.* 1990*a*). Therefore, it seems clear that sex differences in the serum CK response after exercise are, to an important extent, the result of hormonal differences.

1.3.4 Magnetic resonance imaging and spectroscopy

Magnetic resonance (MR) imaging can be used to localize sites of damage, avoiding the need for muscle biopsies. Areas of increased signal intensity on T2–weighted images can be seen immediately after *concentric* exercise (Fleckenstein *et al.* 1988; Shellock *et al.* 1991). These changes in signal intensity are probably due to an increase in intra- and extracellular water content (Fleckenstein *et al.* 1988). *Eccentric* exercise also causes large increases in signal intensity on T1– and T2–weighted images, which occur only 48–96 h afterwards (Fleckenstein *et al.* 1989; Shellock *et al.* 1991). When T1 and T2 relaxation times are calculated from spin echo and inverse recovery images, increases in T1 and T2 are seen (Rodenburg *et al.* 1994*c*). These late increases in signal intensity are probably due to the development of oedema, and in our own study occurred in parallel with the increases in muscle surface area caused by oedema. Increases in signal intensity after eccentric exercise are significantly correlated with the extent of morphological muscle damage (Nurenberg *et al.* 1992). We followed the time course of MR intensity changes and DOMS after eccentric exercise in 8 subjects to investigate whether DOMS could be caused by oedema. It was found that DOMS was maximal at 48 h, at which stage the signal intensity and the muscle surface area had only just started to increase. Signal intensity was highest at 72 and 96 h, and therefore oedema is unlikely to be the cause of DOMS (Rodenburg *et al.* 1994*b*; see also Fig. 1.2). This is in agreement with the results of other authors who used different techniques to measure oedema (Bobbert *et al.* 1986; Clarkson *et al.* 1992). It should be borne in mind that MR imaging cannot provide us with precise information on processes occurring at the cellular or subcellular level, so that it must be used in combination with other techniques if progress is to be made in identifying the underlying causes of muscle damage.

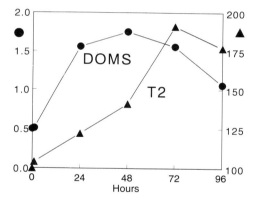

Fig. 1.2 T2 and pain (DOMS) after eccentric exercise in eight subjects. DOMS (filled circles; pain on pressure, left *y*-axis) is given as a pain score on a six-point scale. T2 values (filled triangles, right *y*-axis) are represented as percentage with respect to resting values. The *x*-axis gives the time in hours after exercise, the zero-point being the value before exercise.

The metabolic consequences of damaging exercise can be studied adequately with phosphorus magnetic resonance spectroscopy (^{31}P-MRS). This method shows that there is an increase in the ratio of inorganic phosphate (P_i) to phosphocreatine (PCr) within 1 h of eccentric exercise, and the ratio remains elevated for 72 h before it slowly recovers (Aldridge *et al.* 1986; McCully *et al.* 1988; Rodenburg *et al.* 1994*c*). These findings seem to exclude a purely metabolic explanation of the decline and recovery of force after eccentric exercise: force reaches a minimum value 1 h after exercise, but it recovers steadily during the period when the P_i/PCr ratio is still maximally elevated. It therefore seems more likely that mechanical disruption (over-stretching of sarcomeres, for example) is responsible for the observed decrease in force. Results with ^{31}P-MRS should be interpreted with caution, because of problems related to localization and the homogeneity of the tissue sampled. The total ^{31}P-signal comes from both active and inactive muscle, which influences the P_i/PCr ratio. Moreover, transient changes in P_i/PCr ratio also occur in some muscle disorders (McCully *et al.* 1988).

Although changes in P_i/PCr ratio do not provide a complete explanation for changes in force-generating capacity, the fact that these changes in phosphorus metabolites are present after eccentric exercise may have an important effect on metabolism during subsequent, concentric exercise. We reasoned that the capacity to convert PCr to P_i may be reduced after eccentric exercise, resulting in a reduced performance capacity in a subsequent concentric bout. When this proposition was tested, however, the results suggested that quadriceps muscle metabolism during concentric exercise and recovery is not affected by prior eccentric overload. All the same, the idea that metabolism alters during exercise after a more strenuous prior eccentric overload cannot be ruled out (Rodenburg *et al.* 1995).

1.4   Mechanisms of exercise-induced muscle damage

Two hypotheses have been proposed to explain damage to skeletal muscle resulting from exercise (Armstrong 1984, 1986; Ebbeling and Clarkson 1989). The first hypothesis assumes a metabolic overload, in which the demand for ATP exceeds its production, leading to a vicious cycle of Ca^{2+} overloading of the cell and a further decrease in ATP production. Several observations support the metabolic hypothesis, the most convincing perhaps being the resemblance of exercise-induced muscle damage to ischaemic muscle damage (Armstrong 1986; Hoppeler 1986). Furthermore, there is a group of muscle disorders with a known metabolic defect that leads to exercise intolerance and excessively high CK and Mb after exercise. In extreme cases, rhabdomyolysis with subsequent renal damage may develop in, for example, McArdle's disease and carnitine palmitoyl-transferase deficiency (Brooke 1986). The post-exercise CK increase in patients with chronic progressive external ophthalmoplegia, a mild form of mitochondrial myopathy, is related to the severity of the enzyme defect (Driessen *et al.* 1987). Thus, metabolic factors seem to be important in the pathophysiology of exercise-induced muscle damage.

The fact that eccentric exercise, in particular, induces extensive morphological damage and large increases in CK activity, together with pain, stiffness, and weakness, provides support for the second hypothesis: that mechanical factors are a cause of exercise-induced muscle damage (Newham

1988; Ebbeling and Clarkson 1989). The metabolic cost is lower in eccentric exercise, but the mechanical strain per muscle fibre is higher, as fewer fibres are recruited. Lesions in the banding pattern as well as in the sarcolemma have been observed directly after eccentric exercise (Armstrong *et al.* 1983) and may reflect mechanical damage resulting from high forces on individual fibres. An integrated model of muscle damage has been proposed (Armstrong 1990) that recognizes four stages:

(1) initial events;
(2) autogenetic mechanisms;
(3) phagocytic phase;
(4) regenerative phase.

Initial events, which may be mechanical or metabolic in nature, lead to an autogenetic phase that causes or exacerbates muscle damage via activation of several mechanisms. Calcium overload, caused by the initial event, is generally thought to play a key role in this phase and much attention has been paid to the role of Ca^{2+} and free oxygen radicals (see §6.2.3).

1.5   Drug-induced muscle damage

One possible approach to the study of the mechanisms underlying muscle damage is to investigate the effects of drugs on muscles. There are more than 100 drugs that can impair muscle function (Argov and Mastaglia 1988; Aldrich and Prezant 1990), and some of them cause a myopathy (Kuncl and George 1993). One recent addition to this list is formed by the so-called statins: 3–hydroxy-3–methylglutaryl coenzyme A reductase inhibitors (HMGRIs), a group of cholesterol-lowering agents (Alberts 1988; Goldman *et al.* 1989; Deslypere and Vermeulen 1991; Schalke *et al.* 1992).

In-vivo experiments in our laboratory indicate that rats develop HMGRI-induced muscle damage in a dose-dependent way (Reijneveld *et al.* 1996). The observation that muscle tissue is less sensitive to the more hydrophilic HMGRIs, such as pravastatin (Smith *et al.* 1991; Reijneveld *et al.* 1996), may be explained by differences in tissue selectivity. Lipophilic agents enter the cell easily, whereas the more hydrophilic agents remain extracellular (Serajuddin *et al.* 1991; Koga *et al.* 1992).

The aetiology of this myopathy is poorly understood. It is possible that the cholesterol component of the muscle membrane is affected (Gadbut *et al.* 1995). This may cause an increase in membrane fluidity, resulting in membrane defects that allow influx of Ca^{2+} into muscle fibres. The Ca^{2+} overload would lead to the same sequence of events as has been described for exercise-induced muscle damage. This view is supported by the observation that other hypocholesterolaemic agents with completely different chemical structures, like clofibrate, also cause myopathy (Bregestovski and Bolotina 1989; London *et al.* 1991).

Young, growing rats appear to be more susceptible than adult animals (Smith *et al.* 1991; Reijneveld *et al.* 1996). There are indications that HMGRIs, in addition to their direct toxicity towards muscle tissue, inhibit the proliferation and differentiation of myogenic precursor or satellite cells which takes place in growing and injured muscle (Headley *et al.* 1985; Schultz 1989; Jacobs *et al.* 1995). HMGRIs interfere with the metabolism of mevalonate, a precursor of cholesterol (Goldstein and Brown 1990). Intracellular proteins that play an essential role in the proliferation or differentiation of satellite cells may require covalent binding to farnesol, an intermediate in mevalonate metabolism. Farnesylation may enable these proteins (p21ras, laminin B) to bind to the inner plasma membrane, thus providing signalling mechanisms for mitogenic or myogenic factors (Corsini *et al.* 1992; O'Donnell *et al.* 1993; Masters *et al.* 1995). This could be the reason that muscle tissue is more sensitive to HMGRI-treatment in young, fast growing organisms than in adult animals. We are currently investigating these possibilities *in vitro*.

In conclusion, HMGRI-induced myopathy is not only important from a clinical point of view (Alberts 1988; Goldman *et al.* 1989; Deslypere and Vermeulen 1991; Schalke *et al.* 1992; Contermans *et al.* 1995; Smit *et al.* 1995), but may also serve as an useful *in-vivo* and *in-vitro* model for studying the pathological reactions of muscle and the effects of supplementation with growth and other factors.

1.6   Prevention of exercise-induced muscle damage

1.6.1   Pharmacological studies

1.6.1.1   Calcium

High levels of Ca^{2+} in muscle trigger a series of events leading to muscle damage and necrosis (Wrogeman and Pena 1976; Barracos *et al.* 1986; Soza *et al.* 1986). The underlying mechanism is unknown but may involve activation of a Ca^{2+}-activated neutral protease (Belcastro 1993). Several groups have therefore looked at the possible protective effect of Ca^{2+}-channel antagonists, but with conflicting results (for a review see pp. 83–5 in Amelink 1990). This may be because most of these drugs affect the Ca^{2+} flux over the plasma membrane, whereas skeletal muscle is relatively independent of external Ca^{2+} and relies on Ca^{2+} released from, and taken up by, the sarcoplasmic reticulum. We found that the enzymatic activity of Ca^{2+}-ATPase in the sarcoplasmic reticulum membrane was elevated after exercise, probably through an increased proportion of functional Ca^{2+}-ATPase proteins, which points to the importance of this Ca^{2+}-ATPase in maintaining Ca^{2+}-homeostasis (Ferrington *et al.* 1996). Dantrolene sodium, which interferes with the Ca^{2+} flux across the sarcoplasmic reticulum, does indeed protect rats against muscle damage after running, supporting the hypothesis that Ca^{2+} release from the sarcoplasmic reticulum is an important factor in exercise-induced muscle damage (Amelink *et al.* 1990*b*; Byrd 1992).

1.6.1.2 Free radicals

Exercise causes an increased production of free radicals (Jenkins 1988), which may start a chain reaction leading to focal muscle necrosis. Evidence for the involvement of lipid peroxidation comes from studies in humans (Kanter *et al.* 1986; Maughan *et al.* 1989; Kanter 1994) and in animals (Brady *et al.* 1979; Alessio and Goldfarb 1982). We have demonstrated that vitamin E deficiency increased further the histologically demonstrable muscle damage and raised CK levels produced by 2 h of running (Amelink *et al.* 1991). However, vitamin E supplementation did not prevent muscle damage after eccentric exercise (Warren *et al.* 1992; Jakeman and Maxwell 1993). Several authors have shown that vitamin E deficiency enhances the susceptibility of skeletal muscle to lipid peroxidation (Jackson *et al.* 1983; Salminen *et al.* 1984) and to exercise-induced damage (Jackson 1987). However, Tiidus and Houston (1993) did not find muscle vitamin E concentrations to be reduced in normal or vitamin E-deprived rats directly post-exercise, suggesting that the putative protective role of vitamin E is not due to anti-oxidant effects. Whether vitamin E has membrane-stabilizing effects or attenuates lipid peroxidation, or both, remains to be elucidated. Administration of coenzyme Q_{10}, which is thought to have anti-oxidative and membrane-stabilizing effects, attenuated the exercise-induced serum enzyme increases immediately after eccentric treadmill exercise in rats but had no effect on a second peak 40 h after exercise (Shimomura *et al.* 1991). (For further discussion see §6.2.3.)

1.6.1.3 Oestradiol and tamoxifen

Male-female differences in exercise-induced muscle damage have been attributed to hormonal influences. Recently it has been pointed out that both oestrogen and tamoxifen, a partial oestrogen antagonist, have anti-oxidant as well as membrane-stabilizing properties (Wiseman 1994). It has been shown that tamoxifen reduces CK release in an *in-vitro* model using the soleus muscle (Koot *et al.* 1991). Treatment of male rats with oestradiol resulted in a marked attenuation of the CK response (Bär *et al.* 1988), in parallel with a decrease in the amount of histological damage (Reijneveld *et al.* 1994). Vitamin E deficiency enhances the susceptibility to exercise-induced muscle damage in male rats more than in female rats, probably because of an additional protective effect of oestrogens (Amelink *et al.* 1991). The observation that in female rats, unlike male rats, the vitamin E status does not affect the outcome of training and exercise supports this view (Tiidus and Houston 1993). Taken together these observations suggest that the protective effect of oestrogen may be caused by its free-radical scavenging properties, rather than by hormonal activity (Tiidus 1995).

1.6.1.4 Corticosteroids

It has been demonstrated in several clinical trials that prednisone, a glucocorticoid, improves strength in patients with Duchenne muscular dystrophy (DMD). The hypothesis was advanced that this effect was attributable not to immunosuppressive mechanisms but to an increased stability of the muscle fibre membrane (Brooke *et al.* 1987; Mendell *et al.* 1989; Griggs *et al.* 1993; Kissel *et al.* 1993). This stabilizing effect could be due to an increased expression of dystrophin (Hardiman *et al.* 1993), or an increased synthesis of lipocortin resulting in decreased phospholipase activity (Flower 1990). Exercise causes the serum activity of several proteins to rise, reflecting an increased release or leakage from muscle that may be due to disruption of the muscle fibre membrane (Bär *et al.* 1990; Bär and Amelink 1992). On the basis that the reaction of healthy muscle to exercise may be similar to the reaction of dystrophin-deficient muscle in DMD patients to the strain of normal daily activity, we tested in rats whether prednisone would protect normal muscle fibres against mechanically induced muscle damage. We did indeed find a dose-related protective effect of prednisone on exercise-induced histological muscle damage. Prednisone did not, however, prevent leakage of CK from the muscle immediately after exercise, nor did it induce increased expression of dystrophin or stimulate the proliferative activity of satellite cells. Thus, the mechanism by which steroids improve strength in dystrophin-deficient muscle and prevent exercise-induced damage in normal muscle remains a puzzle (Jacobs *et al.* 1995).

1.6.2 Training

It is well established that muscle damage can be prevented by training, whether it involves concentric (Bosman *et al.* 1993) or eccentric exercise (Clarkson and Tremblay 1988; Balnave and Thompson 1993). The protective effect of a single strenuous bout of eccentric exercise may last up to months. The changes within the muscle that are responsible for this adaptation are largely unknown. Structural as well as metabolic changes have been mentioned (Ebbeling and Clarkson 1989; Nosaka and Clarkson 1992; Kuipers 1994). It has been suggested that a group of more fragile, stress-susceptible fibres exists; these would be reduced in number after a first bout of exercise while the stronger fibres survived (Armstrong *et al.* 1983; Morton and Carter 1992). However, even light eccentric training protocols that do not lead to an increase in serum CK activity are still sufficient to bring about a protective effect after a second bout of training (Clarkson and Tremblay 1988). It may be that part of the effect lies outside the muscle. The central nervous system (CNS) can drastically adapt its output after a single training session, especially when the conditions are very alarming or damaging (cf. the passive avoidance response in behavioural studies (Ader *et al.* 1972)). Thus the eccentric exercise session that leads to serious muscle injury (pain, loss of function and force, muscle damage) may be regarded as a learning experience to which the CNS responds, for example, by adapting the way in which it employs motor units during similar exercises long afterwards (Wernig *et al.* 1991; Clarkson *et al.* 1992). This explanation is hypothetical and remains to be tested experimentally. It is, however, supported by the observation that adaptation occurs soon after the first bout of exercise, at a time that the muscle is still 'damaged' or recovering from the insult (Ebbeling and Clarkson 1990; Bosman *et al.* 1993). When CK was measured in rats before and after an initial training run, and then again after a second run 1, 2, 3, or 4 days later, it appeared that adaptation occurred between 24 and 48 h after the first run. Both CK release and morphological damage were significantly reduced (Fig. 1.3; Bär and Reijneveld 1993). It should be noted that level running results in less damage and may therefore have a smaller 'learning' effect than eccentric exercise (Ader *et al.* 1972).

1.6.3   Warming up, stretching, and massage

The functional changes that accompany morphological damage (decreased maximal force and range of motion, DOMS) hinder optimal performance in sports activities. Measures such as warm-up, stretching exercises, and massage are often taken in an attempt to avoid these undesirable consequences of exercise. The effect of these measures has not been studied to any extent. De Vries (1960) found that stretching exercises had an ameliorative effect on the development of DOMS after eccentric exercise, but this finding was contradicted by other workers (McGlynn *et al.* 1979; High and Howley 1989). The conflict between these studies may have been the result of the subjective nature of the DOMS score: whereas the scores of a single subject can be compared adequately on successive days, the scores of different subjects are hard to compare, since different individuals perceive pain differently. DOMS scores are poorly related to other outcomes of exercise (Rodenburg *et al.* 1993) and to the amount of morphological damage (Nurenberg *et al.* 1992). Until now objective studies on functional and biochemical changes caused by eccentric exercise have not produced evidence that interventions are actually useful: stretching after eccentric exercise did not affect DOMS, CK release, or decrease in force (Buroker and Schwane 1989), nor did massage after eccentric exercise (Wenos *et al.* 1990).

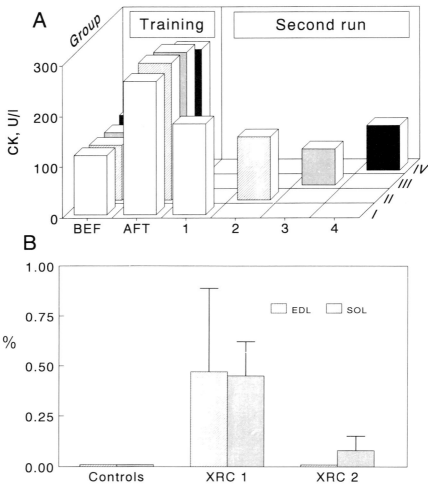

Fig. 1.3 (A) The effects of one training session on creatine kinase (CK) release after a second run. Four groups (I, II, III, IV) were trained at day 1 for 2 h on a treadmill. The CK activity in serum was measured before (BEF) and immediately after (AFT) running. Then each group ran a second time: group I the next day (1); group II 2 days later (2); group III 3 days later (3); and group IV 4 days later (4). The CK value after exercise was determined and is given in the figure. The differences between the first CK value and the second are significant for groups II, III, and IV. (B) Quantification of muscle damage. The percentage damaged area (*y*-axis) is given for soleus (SOL) and extensor digitorum longus (EDL). Both muscles were taken out 48 h after the first (XRC 1) and second (XRC 2) exercise session of group IV. Controls are animals that were treated identically except that they did not run.

We investigated recently whether a combined intervention—warm-up and stretching before exercise, and massage afterwards—would diminish CK and Mb efflux, maximal isotonic force, range of motion of the elbow, and DOMS resulting from eccentric arm exercise. All parameters were less affected in the treated group, and the differences in CK efflux, flexion angle, and maximal isotonic force were significant. Since all changes were in the same direction and three of them were significant, it was concluded that a combined intervention could indeed reduce changes caused by damaging eccentric exercise (Rodenburg *et al.* 1992*c*, 1994*a*). Whether this effect is attributable to changes in muscular blood flow, neural activation, elasticity of the muscle, or other, as yet unknown, mechanisms remains to be established. (See §9.3.7 for a further discussion of this issue.)

1.7   Regeneration and repair

Skeletal muscle has an enormous capacity for regeneration, thanks to the presence of satellite cells. Under normal conditions these cells are quiescent in adult muscle, where they may be regarded as reserve stem cells (Mauro 1961). They are activated by damaging influences and then proliferate (Schultz 1989; Grounds 1991). When the basal lamina of the damaged muscle is intact, satellite cells form myogenic cells which fuse with existing fibres (moderate injury) or with each other (extensive damage) to form myotubes (White and Esser 1989). Training and muscle activity also stimulate satellite cells: hypertrophy and hyperplasia, both of which would lead to increased muscle mass, are believed to be based on activation of satellite cells (Darr and Schulz 1987). Development, hypertrophy, and regeneration may be considered points in a continuum with respect to cell growth (White and Esser 1989). Conversely, atrophy resulting from decreased use is associated with decreased mitotic activity.

Damage to the muscle may act as a stimulus for muscle development or repair (Jones *et al.* 1989). *In-vitro* experiments (Bischoff 1989) have shown that extracts of injured muscles stimulate the proliferation of satellite cells, the effect reaching a peak after 8–12 h. Continuous exposure to such extracts was necessary to sustain proliferation and it appeared that, whereas factors from the extract are probably essential for stimulating the quiescent satellite cell to proliferate, other factors are needed to maintain progression through the cell cycle.

In-vivo studies with rats of two different age groups (6 and 16 weeks), killed at different times after a 2 h run on a level treadmill, showed that damage and satellite cell activation develop quickly. Histological evidence of damage differed in both amount and time course between young and adult rats. In young rats the amount of damage reached a maximum at 48 h (8 per cent of total area studied) and subsided within 1 week of the exercise. In the older animals the damage was lower in amount (a maximum of 4 per cent 1 week after exercise) but remained present for at least 2 weeks. The rats were injected with bromodeoxyuridine 2 h before they were killed, to study the dynamics of satellite cell proliferation. In young rats the number of satellite cells per cross-section of soleus increased 3– to 4–fold between 12 and 24 h after exercise and

Muscle Damage

remained elevated thereafter. The labelling index, a measure of proliferative activity, increased from 24 to 34 per cent in the same period. In adult rats the total number of satellite cells increased between 48 h and 2 weeks after exercise. The labelling index increased from 0 to 23 per cent after 1 week (Jacobs *et al.* 1993, 1995; Fig. 1.4). This level running exercise did not lead to widespread necrotic changes in the muscles studied, yet—as others have found (Darr and Schulz 1987)—satellite cells proliferated throughout the muscle, and not only in or around foci of damage. This suggests that activation is mediated by diffusible factors generated by exercise (Florini and Magri 1989). The rapidity and extent of the proliferative response could be due to leakage of mitogenic factors through small membrane disruptions. It is reported that basic fibroblast growth factor (bFGF) release could be coupled to plasma membrane wounding (Clarke *et al.* 1993; McNeil 1993) and that damaged muscle fibres release other mitogenic factors (Bischoff 1990).

Several growth factors stimulate or inhibit proliferation and differentiation of satellite cells. The factors that have received most attention are bFGF, insulin-like growth factor I (IGF-I), and transforming growth factor-beta (TGF-β) (White and Esser 1989; Allen and Rankin 1990; Guthridge *et al.* 1992). These factors are involved during muscle growth in early life as well as in hypertrophy and regeneration. They can be made available to muscle in various ways: bFGF is bound to a component of the extracellular matrix and is thus contained in the same compartment as the satellite cell (Yamada *et al.* 1989). Disruption of the extracellular matrix after eccentric exercise (Stauber *et al.* 1990) could cause bFGF to be released. bFGF is assumed to be involved as a mitogen mainly in the early phases of regeneration (Allen and Rankin 1990). Macrophages that enter the muscle 24–48 h after exercise (Smith 1991) to remove necrotic debris are a further potential source of bFGF (Allen and Rankin 1990). IGF-I is synthesized in satellite cells (Jennische *et al.* 1987) but is also present in the circulation where it is bound to several proteins (Nissley and Rechler 1984). IGF-I has only a slight effect on proliferation, but stimulates differentiation very strongly *in vitro*. Finally, TGF-β inhibits differentiation and might play a role when proliferation is more important than differentiation (in early regeneration) and when both differentiation and proliferation need to be repressed (in the quiescent state).

Together, these factors could control the entire cycle of the satellite cell (Allen and Rankin 1990). It should be kept in mind, however, that most evidence has been gathered from *in-vitro* systems, and that discrepancies have been found between *in-vitro* and *in-vivo* behaviour (Guthridge *et al.* 1992). It is very likely that other hormones and factors, such as ciliary neurotrophic factor (CNTF, DiStefano *et al.* 1992), retinoic acid, and thyroid hormone (Sugie and Verity 1985; D'Albis *et al.* 1990), also contribute to the control of myogenesis. We are currently investigating the effects of these factors on muscle satellite cells in an *in-vitro* model (Jacobs *et al.* 1996 a, b and manuscript submitted for publication).

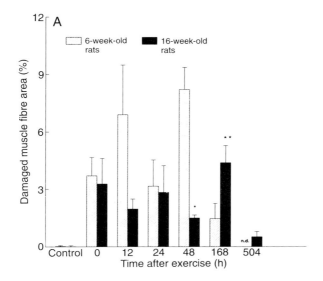

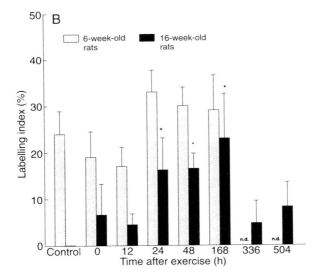

Fig. 1.4 (A) Damaged muscle fibre area (%) in longitudinal sections in the proximal part of the soleus muscle at different times after exercise. The values are expressed as mean ± SEM. Control, without exercise. All values, except the value at 504 h (16–week-old rats), are significantly different from the control. *, Significantly lower than in the younger animals; **, significantly higher than in the younger animals ($p \leqslant 0.05$); n.d., not done. (B) Time course of the labelling index of satellite cell nuclei after exercise in 6– and 16–week-old animals. *, Significantly different from the control group ($p \leqslant 0.05$); n.d., not done.

New approaches to the study of muscle repair will certainly develop from studies on factors that regulate the expression of muscle-specific genes, such as MyoD, Myf5, Myf6, and myogenin (Russell *et al.* 1992); these genes are expressed only in developing and regenerating muscle. We are only beginning to understand the timing of the expression of these factors. Certain non-contractile proteins, members of the family of intermediate filaments (desmin, vimentin) and the large protein titin, play an important role in normal function but may also be involved in the structural organization that forms part of the early stages of regeneration. Although there are again differences between the behaviour of these proteins *in vivo* and in tissue culture, they have promise as markers for regenerating muscle (Bornemann and Schmalbruch 1992; Vater *et al.* 1992; see also Chapter 10). Unravelling the control of muscle repair may lead to new approaches to the treatment of muscle damage resulting from disease.

1.8   Conclusions

It is of theoretical and practical importance to know how the different indicators for muscle damage are related. Does a large CK efflux correspond to a large extent of morphological damage and large decreases in force, or do the indicators change independently? When we investigated the relationship between CK, Mb, range of motion, and force (Rodenburg *et al.* 1993), we found little correlation between them, implying that they are separate components of a process to which changes in the different variables are related. In the same study, we found hardly any correlation between these indicators and DOMS. This may be explained in part by the subjective nature of a DOMS score. Of the phenomena described in this paper, post-exercise pain, in general, remains the greatest enigma. To prevent pain and other unwanted side-effects of exercise, it seems that training is still the best option (Bär *et al.* 1990), combined if necessary with warming-up, stretching, and massage. With respect to fundamental insights into damage and repair, progress can be anticipated in the near future from studies of the trophic factors involved, and the way in which their effects are expressed via muscle-specific genes. These genes, and their products, can be expected to be better markers for damage and repair than the indicators described in this chapter. Drug-induced muscle damage could provide a useful model for such studies.

Acknowledgements

The work described in this chapter has been performed over the last 10 years by a number of people, both scientists and volunteers. The latter will remain anonymous but are immensely important. The former are acknowledged through their co-authorship of publications cited in this review.

References

Ader, R., Weijnen, J.A.W.M., and Moleman, P. (1972). Retention of a passive avoidance response as a function of the intensity and duration of electric shock. *Psychonomic Science*, **26**, 125–8.

Alberts, A.W. (1988). HMG-CoA reductase inhibitors—the development. *Atherosclerosis Reviews*, **18**, 123–31.

Aldrich, T.K. and Prezant, D.J. (1990). Adverse effects of drugs on the respiratory muscles. *Clinics in Chest Medicine*, **11**, 177–89.

Aldridge, R., Cady, E.B., Jones, D.A., and Obletter, G. (1986). Muscle pain after exercise is linked with an inorganic phosphate increase as shown by ^{31}P NMR. *Bioscience Reports*, **6**, 663–7.

Alessio, H.M. and Goldfarb, A.H. (1982). Lipid peroxidation and scavenger enzymes during exercise: adaptive response to training. *Journal of Applied Physiology*, **64**, 1333–6.

Allen, R.E. and Rankin, L.L. (1990). Regulation of satellite cells during skeletal muscle growth and development. *Proceedings of the Society for Experimental Biology and Medicine*, **194**, 81–6.

Amelink, G.J. (1990). *Exercise-induced muscle damage*. Ph.D. Thesis. Krips Repro, Meppel, The Netherlands.

Amelink, G.J. and Bär, P.R. (1986). Exercise-induced muscle protein leakage in the rat: effects of hormonal manipulation. *Journal of the Neurological Sciences*, **76**, 61–8.

Amelink, G.J., Kamp, H.H., and Bär, P.R. (1988). Creatine kinase isoenzyme profiles after exercise in the rat. Sex-linked differences in leakage of CK-MM. *Pflügers Archiv—European Journal of Physiology*, **412**, 417–21.

Amelink, G.J., Koot, R., Erich, W.B.M., van Gijn, J., and Bär, P.R. (1990*a*). Sex-linked variation in creatine kinase release, and its dependence on oestradiol, can be demonstrated in an in vitro rat muscle preparation. *Acta Physiologica Scandinavica*, **138**, 115–24.

Amelink, G.J., van der Kallen, C.J.H., Wokke, J.H.J., and Bär P.R. (1990*b*). Dantrolene sodium diminishes exercise-induced muscle damage in the rat. *European Journal of Pharmacology*, **179**, 187–92.

Amelink, G.J., van der Wal, W.A.A., van Asbeck, B.S., and Bär, P.R. (1991). Exercise-induced muscle damage in the rat: the effect of vitamin E deficiency. *Pflügers Archiv—European Journal of Physiology*, **419**, 304–9.

Argov, Z. and Mastaglia, F.L. (1988). Drug-induced neurological disorders in man. In *Disorders of voluntary muscle* (ed. J. Walton), pp. 981–1014. Churchill-Livingstone, Edinburgh.

Armstrong, R.B. (1984). Mechanisms of exercise-induced delayed onset muscular soreness: a brief review. *Medicine and Science in Sports and Exercise*, **16**, 529–38.

Armstrong, R.B. (1986). Muscle damage and endurance events. *Sports Medicine*, **3**, 370–81.

Armstrong, R.B. (1990). Initial events in exercise-induced muscular injury. *Medicine and Science in Sports and Exercise*, **22**, 429–35.

Armstrong, R.B., Ogilvie, R.W., and Schwane, J.A. (1983). Eccentric exercise-induced injury to rat skeletal muscle. *Journal of Applied Physiology: Respiratory, Environmental, and Exercise Physiology*, **54**, 80–93.

Balnave, C.D. and Thompson, M.W. (1993). Effect of training on eccentric exercise-induced muscle damage. *Journal of Applied Physiology*, **75**, 1545–51.

Bär, P.R. and Amelink, G.J. (1992). Creatine kinase and its isoenzymes as indices for muscle damage. In *Guanidino compounds in biology and medicine* (ed. P.P. de Deyn, B. Marescau, V. Stalon, and I.A. Qureshi), pp. 231–8. Libbey & Company, London.

Bär, P.R. and Reijneveld, J.C. (1993). The effect of a single training session on exercise-induced muscle damage. *Medicine and Science in Sports and Exercise*, **25** (Suppl.), S72.

Bär, P.R., Amelink, G.J., and Blankenstein, M.A. (1988). Prevention of exercise-induced muscle membrane damage by oestrogen. *Life Sciences*, **42**, 2677–81.

Bär, P.R., Amelink, G.J., Jackson, M.J., Jones, D.A., and Bast, A. (1990). Aspects of exercise-induced muscle damage. In *Sports, medicine and health* (ed. G.P.H. Hermans), pp. 1143–8. Elsevier Science Publishers, Amsterdam.

Barracos, V., Greenberg, R.E., and Goldberg, A.L. (1986). Influence of calcium and other divalent cations on protein turnover in rat skeletal muscle. *American Journal of Physiology*, **250**, E702–10.

Belcastro, A.N. (1993). Skeletal muscle calcium-activated neutral protease (calpain) with exercise. *Journal of Applied Physiology*, **74**, 1381–6.

Berg, A. and Keul, J. (1981). Physiological and metabolic responses of female athletes during laboratory and field exercise. *Medicine Sport*, **14**, 77–96.

Berry, C.B., Moritani, T., and Tolson, H. (1990). Electrical activity and soreness in muscles after exercise. *American Journal of Physical Medicine and Rehabilitation*, **69**, 60–6.

Bischoff, R. (1989). Analysis of muscle regeneration using single myofibers in culture. *Medicine and Science in Sports and Exercise*, **21**, S164–72.

Bischoff, R. (1990). Interaction between satellite cells and skeletal muscle fibers. *Development*, **109**, 943–52.

Bobbert, M.F., Hollander, A.P., and Huijing, P.A. (1986). Factors in delayed onset muscular soreness of man. *Medicine and Science in Sports and Exercise*, **18**, 75–81.

Borleffs, J.C.C., Derksen, R.H.W.M., and Bär, P.R. (1987). Serum myoglobin and creatine kinase concentrations in patients with polymyositis or dermatomyositis. *Annals of the Rheumatic Diseases*, **46**, 173–5.

Bornemann, A. and Schmalbruch, H. (1992). Desmin and vimentin in regenerating muscles. *Muscle and Nerve*, **15**, 14–20.

Bosman, P.J., Balemans, W.A.F., Amelink, G.J., and Bär, P.R. (1993). A single training session affects exercise-induced muscle damage in the rat. In *Neuromuscular fatigue* (ed.A.J. Sargeant and D. Kernell), pp. 74–5. North-Holland, Amsterdam.

Brady, P.S., Brady, L.J., and Ullrey, D.E. (1979). Selenium, vitamin E and the response to swimming stress in the rat. *Journal of Nutrition*, **109**, 1103–9.

Bregestovski, P.D. and Bolotina, V.N. (1989). Membrane fluidity and kinetics of Ca^{2+}-dependent potassium channels. *Biomedica et Biochimica Acta*, **48**, 382S-7S.

Brooke, M.H. (1986). *A clinician's view of neuromuscular diseases*, 2nd edition. Williams and Wilkins, Baltimore.

Brooke, M.H., Fenichel, G.M., Griggs, R.C., Mendell, J.R., Moxley, R.T., Miller, J.P. *et al.* (1987). Clinical investigation of Duchenne muscular dystrophy. Interesting results in a trial of prednisone. *Archives of Neurology*, **44**, 812–17.

Buroker, K.C. and Schwane, J.A. (1989). Does postexercise static stretching alleviate delayed muscle soreness? *Physician and Sportsmedicine*, **17**, 65–83.

Byrd, S.K. (1992). Alterations in the sarcoplasmic reticulum. a possible link to exercise-induced muscle damage. *Medicine and Science in Sports and Exercise*, **24**, 531–6.

Clarke, M.S.F., Khakee, R., and McNeil, P.L. (1993). Loss of cytoplasmic basic fibroblast growth factor from physiologically wounded myofibers of normal and dystrophic muscle. *Journal of Cell Science*, **106**, 121–33.

Clarkson, P.M. and Tremblay, I. (1988). Exercise-induced muscle damage, repair, and adaptation in humans. *Journal of Applied Physiology*, **65**, 1–6.

Clarkson, P.M., Nosaka, K., and Braun, B. (1992). Muscle function after exercise-induced muscle damage and rapid adaptaion. *Medicine and Science in Sports and Exercise*, **24**, 512–20.

Contermans, J., Smit, J.W., Bär, P.R., and Erkelens, D.W. (1995). A comparison of the effects of simvastatin and pravastatin monotherapy on muscle histology and permeability in hypercholesterolaemic patients. *British Journal of Clinical Pharmacology*, **39**, 135–41.

Corsini, A., Raiteri, M., Soma, M.R., Gabbiani, G., and Paoletti, R. (1992). Simvastatin but not pravastatin has a direct inhibitory effect on rat and human myocyte proliferation. *Clinical Biochemistry*, **25**, 399–400.

D'Albis, A., Chanoine, C., Janmot, C., Mira, J.C., and Couteaux, R. (1990). Muscle specific response to thyroid hormone of myosin isoform transitions during postnatal development. *European Journal of Biochemistry*, **93**, 155–61.

Darr, K.C. and Schulz, E. (1987). Exercise-induced satellite cell activation in growing and mature skeletal muscle. *Journal of Applied Physiology*, **63**, 1816–21.

Dempsey, R., Morgan, J., and Cohen, L. (1975). Reduction of enzyme efflux from skeletal muscle by diethylstilbestrol. *Clinical Pharmacology and Therapeutics*, **18**, 104–11.

Deslypere, J.P. and Vermeulen, A. (1991). Rhabdomyolysis and simvastatin. *Annals of Internal Medicine*, **114**, 342.

DeVries, H.A. (1960). Prevention of muscular distress after exercise. *Research Quarterly for Exercise and Sports*, **2**, 177–85.

DiStefano, P.S., Yancopoulos, G.D., and Squinto, S.P. (1992). Ciliary neurotrophic factor (CNTF) prevents denervation-induced atrophy of rat skeletal muscle. *Neurology*, **42** (S), 669P.

Donnely, A.E., Maughan, R.J., and Whiting, P.H. (1990). Effect of ibuprofen on exercise-induced muscle soreness and indices of muscle damage. *British Journal of Sports Medicine*, **24**, 191–5.

Driessen, M.F., Bär, P.R., Scholte, H.R., Hoogenraad, T.U., and Luyt-Houwen, I.E.M. (1987). A striking correlation between muscle damage after exercise and mitochondrial dysfunction in patients with chronic progressive external ophthalmoplegia. *Journal of Inherited Metabolic Diseases*, **10**, 252–5.

Driessen-Kletter, M.F., Amelink, G.J., Bär, P.R., and van Gijn, J. (1990). Myoglobin is a sensitive marker of increased muscle membrane vulnerability. *Journal of Neurology*, **237**, 234–8.

Ebbeling, C.B. and Clarkson, P.M. (1989). Exercise-induced muscle damage and adaptation. *Sports Medicine*, **7**, 207–34.

Ebbeling, C.B. and Clarkson, P.M. (1990). Muscle adaptation prior to recovery following eccentric exercise. *European Journal of Applied Physiology*, **60**, 26–31.

Edwards, R.H.T., Hill, D.K., Jones, D.A., and Merton, P.A. (1977). Fatigue of long duration in human skeletal muscle after exercise. *Journal of Physiology*, **272**, 769–78.

Ferrington, D.A., Rutledge, R.A., Schneider, C.M., and Apple, F.S. (1992). Effects of gender and estrogen administration on muscle CK activity in rat skeletal muscle. *Abstracts 39th Annual Meeting of the American College of Sports Medicine*, Dallas, p. 55.

Ferrington, D.A., Reijneveld, J.C., Bär, P.R., and Bigelow, D.J. (1996). Activation of the sarcoplasmic reticulum Ca-ATPase induced by exercise. *Biochimica et Biophysica Acta*, **1279**, 203–13.

Fleckenstein, J.L., Canby, R.C., Parkey, R.W., and Peshock, R.M. (1988). Acute effects of exercise on MR imaging of skeletal muscle in normal volunteers. *American Journal of Radiology*, **151**, 231–7.

Fleckenstein, J.L., Weatherall, P.T., Parkey, R.W., Payne, J.A., and Peshock, R.M. (1989). Sports related muscle injuries: evaluation with MR imaging. *Radiology*, **172**, 793–8.

Florini, J.R. and Magri, K.A. (1989). Effects of growth factors on myogenic differentiation. *American Journal of Physiology*, **256**, C701.

Flower, R.J. (1990). Lipocortin. In *Cytokines and lipocortines in inflammation and differentiation*, (ed. M. Meli and L. Parente), *Progress in Clinical and Biological Research*, **349**, 11–25. Wiley-Liss, New York.

Fridén, J. and Lieber, R.L. (1992). Structural and mechanical basis of exercise-induced muscle injury. *Medicine and Science in Sports and Exercise*, **24**, 521–30.

Gadbut, A.P., Caruso, A.P., and Galper, J.B. (1995). Differential sensitivity of C2–C12 striated muscle cells to lovastatin and pravastatin. *Journal of Molecular and Cellular Cardiology*, **27**, 2397–402.

Goldman, J.A., Fishman, A.B., Lee, J.E., and Johnson, R.J. (1989). The role of cholesterol-lowering agents in drug-induced rhabdomyolysis and polymyositis. *Arthritis and Rheumatism*, **32**, 358–9.

Goldstein, J.L. and Brown, M.S. (1990). Regulation of the mevalonate pathway. *Nature (London)*, **343**, 425–30.

Griggs, R.C., Moxley, R.T III., Mendell, J.R., Fenichel, G.M., Brooke, M.H., Pestronk, A. *et al.* (1993). Duchenne dystrophy. Randomized, controlled trial of prednisone (18 months) and azathioprine (12 months). *Neurology*, **43**, 520–7.

Grounds, M.D. (1991). Towards understanding skeletal muscle regeneration. *Pathology, Research and Practice*, **187**, 1–22.

Guthridge, M., Wilson, M., Cowling, J., Bertolini, J., and Hearn, M.T.W. (1992). The role of basic fibroblast growth factor in skeletal muscle regeneration. *Growth factors*, **6**, 53–63.

Hardiman, O., Sklar, R.M., and Brown, R.H. (1993). Methylprednisolone selectively affects dystrophin expression in human muscle cultures. *Neurology*, **43**, 342–5.

Hasson, S.M., Daniels, J.C., Divine, J.G., Niebuhr, B.R., Richmond, S., Stein, P.G. *et al.* (1993). Effect of ibuprofen use on muscle soreness, damage, and performance. A preliminary investigation. *Medicine and Science in Sports and Exercise*, **25**, 9–17.

Headley, S.A.E., Newham, D.J., and Jones, D.A. (1985). The effect of prednisolone on exercise-induced muscle soreness and damage. *Clinical Science*, **70**, 85P.

High, D.M. and Howley, E.T. (1989). The effects of static stretching and warm-up on prevention of delayed onset muscle soreness. *Research Quarterly for Exercise and Sport*, **60**, 357–61.

Hoppeler, H. (1986). Exercise-induced ultrastructural changes in skeletal muscle. *International Journal of Sports Medicine*, **7**, 187–204.

Hortobágyi, T. and Denahan, T. (1989). Variability in creatine kinase. Methodological, exercise and clinically related factors. *International Journal of Sports Medicine*, **10**, 69–80.

Howell, J.N., Chila, A.G., Ford, G., David, D., and Gates, T. (1985). An electromyographic study of elbow motion during postexercise muscle soreness. *Journal of Applied Physiology*, **58**, 1713–18.

Jackson, M.J. (1987). Muscle damage during exercise. Possible role of free radicals and protective effect of vitamin E. *Proceedings of the Nutrition Society*, **46**, 77–80.

Jackson, M.J., Jones, D.A., and Edwards, R.H.T. (1983). Lipid peroxidation of skeletal muscle. An in vitro study. *Bioscience Reports*, **3**, 609–19.

Jacobs, S., Bootsma, A.L., Wokke, J.H.J., and Bär, P.R. (1993). Satellite cell activation and muscle damage after exercise in rats. A morphological study. *Medicine and Science in Sports and Exercise*, **25**, S72.

Jacobs, S.J., Wokke, J.H.J., Bär, P.R., and Bootsma, A.L. (1995). Satellite cell activation and muscle damage in young and adult rats. *The Anatomical Record*, **242**, 329–36.

Jacobs, S.C.J.M., Bär, P.R., and Bootsma, A.L. (1996a). Effect of hypothyroidism on satellite cells and postnatal fiber development in the soleus muscle of the rat. *Cell and Tissue Research*, in press.

Jacobs, S.C.J.M., Bootsma, A.L., Willems, P.W.A., Bär, P.R., and Wokke, J.H.J. (1996b). Prednisone can protect against exercise-induced muscle damage. *Journal of Neurology*, **243**, 410–16.

Jakeman, P. and Maxwell, S. (1993). Effect of antioxidant vitamin supplementation on muscle function after eccentric exercise. *European Journal of Applied Physiology*, **67**, 426–30.

Jenkins, R.R. (1988). Free radical chemistry. Relationship to exercise. *Sports Medicine*, **5**, 156–70.

Jennische, E., Skottner, A., and Hansson, H.A. (1987). Satellite cells express the trophic factor IGF-I in regenerating skeletal muscle. *Acta Physiologica Scandinavica*, **129**, 9–15.

Jones, D.A., Newham, D.J., and Clarkson, P.M. (1987). Skeletal muscle stiffness and pain following eccentric exercise of the elbow flexors. *Pain*, **30**, 233–42.

Jones, D.A., Rutherford, O.M., and Parker, D.F. (1989). Physiological changes in skeletal muscle as a result of strength training. *Quarterly Journal of Experimental Physiology*, **74**, 233–56.

Kagen, L.J., Moussavi, S., Miller, S.L., and Tsairis, P. (1980). Serum myoglobin in muscular dystrophy. *Muscle and Nerve*, **3**, 221–6.

Kanter, M.M. (1994). Free radicals, exercise, and antioxidant supplementation. *International Journal of Sports Nutrition*, **4**, 205–20.

Kanter, M.M., Kaminsky, L.A., La Ham-Saeger, J., Lesmes, G.R., and Nequin, N.D. (1986). Serum enzyme levels and lipid peroxidation in ultramarathon runners. *Annals of Sports Medicine*, **3**, 39–41.

Kissel, J.T., Lynn, D.J., Rammohan, K.W., Klein, J.P., Griggs, R.C., Moxley, R.T.

et al. (1993). Mononuclear cell analysis of muscle biopsies in prednisone- and azathioprine-treated Duchenne muscular dystrophy. *Neurology*, **43**, 532–6.

Koga, T., Fukuda, K., Shimada, Y., Fukami, M., and Tsujita, Y. (1992). Tissue selectivity of pravastatin sodium, lovastatin and simvastatin. The relationship between inhibition of de novo sterol synthesis and active drug concentrations in the liver, spleen and testis in rat. *European Journal of Biochemistry*, **209**, 315–19.

Komulainen, J., Kytola, J., and Vihko, V. (1994). Running-induced muscle injury and myocellular enzyme release in rats. *Journal of Applied Physiology*, **77**, 2299–304.

Komulainen, J., Takala, T.E.S., and Vihko, V. (1995). Does increased serum creatine kinase activity reflect exercise-induced muscle damage in rats? *International Journal of Sports Medicine*, **16**, 150–4.

Koot, R.W., Amelink, G.J., Blankenstein, M.A., and Bär, P.R. (1991). Tamoxifen and oestrogen both protect the rat muscle membrane against physiological damage. *Journal of Steroid Biochemistry and Molecular Biology*, **40**, 689–95.

Kuipers, H. (1994). Exercise-induced muscle damage. *International Journal of Sports Medicine*, **15**, 132–5.

Kuipers, H., Keizer, H.A., Verstappen, F.T.J., and Costill, D.L. (1985). Influence of a prostaglandin-inhibiting drug on muscle soreness after eccentric work. *International Journal of Sports Medicine*, **6**, 336–9.

Kuncl, R.W. and George, E.B. (1993). Toxic neuropathies and myopathies. *Current Opinion in Neurology*, **6**, 695–704.

London, S.F., Gros, K.F., and Ringel, S.P. (1991). Cholesterol-lowering agent myopathy (CLAM). *Neurology*, **41**, 1159–60.

Masters, B.A., Palmoski, M.J., Flint, O.P., Gregg,R.E., Wang-Iverson, D., and Durham, S.K. (1995). In vitro myotoxicity of the 3–hydroxy-3–methylglutaryl coenzyme A reductase inhibitors pravastatin, lovastatin and simvastatin, using neonatal rat skeletal myocytes. *Toxicology and Applied Pharmacology*, **131**, 163–74.

Maughan, R.J., Donnelly, A.E., Gleeson, M., Whiting, P.H., Walker, K.A., and Clough, P.J. (1989). Delayed-onset muscle damage and lipid peroxidation in man after a downhill run. *Muscle and Nerve*, **12**, 332–6.

Mauro, A. (1961). Satellite cells of skeletal muscle fibers. *Journal of Biophysical and Biochemical Cytology*, **9**, 493–5.

McCully, K.K., Argov, Z., Boden, B.P., Brown, R.L., Bank, W.J., and Chance, B. (1988). Detection of muscle injury in humans with ^{31}P magnetic resonance spectroscopy. *Muscle and Nerve*, **11**, 212–16.

McGlynn, G.H., Laughlin, N.T., and Rowe, V. (1979). Effect of electromyographic feedback and static stretching on artificially induced muscle soreness. *American Journal of Physical Medicine*, **58**, 139–48.

McNeil, P.L. (1993). Cellular and molecular adaptations to injurious mechanical stress. *Trends in Cellular Biology*, **3**, 302–7.

Meltzer, H.Y. (1971). Factors affecting serum creatine phosphokinase levels in the general population. The role of race, activity and sex. *Clinica Chimica Acta*, **33**, 165–72.

Mendell, J.R., Moxley, R.T., Griggs, R.C., Brooke, M.H., Fenichel, G.M., Miller, J.P. *et al.* (1989). Randomized, double-blind six-month trial of prednisone in Duchenne's muscular dystrophy. *New England Journal of Medicine*, **320**, 1592–7.

Meulen, J. van der (1991). Exercise-induced muscle damage. Thesis, University of Maastricht, The Netherlands.

Miles, M.P. and Clarkson, P.M. (1994). Exercise-induced muscle pain, soreness, and cramps. *Journal of Sports Medicine and Physical Fitness*, **34**, 203–16.

Morton, R.H. and Carter, M.R. (1992). Elevated serum enzyme activity. An explanation-based model. *Journal of Applied Physiology*, **73**, 2192–200.

Newham, D.J. (1988). The consequences of eccentric contractions and their relationship to delayed onset muscle soreness. *European Journal of Applied Physiology*, **47**, 353–9.

Newham, D.J., Mills, K.R., Quigley, B.M., and Edwards, R.H.T. (1983). Pain and fatigue after concentric and eccentric muscle contractions. *Clinical Science*, **64**, 55–62.

Nissley, S.P. and Rechler, M.M. (1984). Insulin-like growth factors. Biosynthesis, receptors and carrier proteins. In *Hormonal proteins and peptides* (ed. C.H. Li), pp. 127–203. Academic Press, New York.

Noakes, T.D. (1987). Effects of exercise on serum enzyme activities in humans. *Sports Medicine*, **4**, 245–67.

Nørregaard-Hansen, K., Bjerre-Knudsen, J., Brodthagen, U., Jordal, R., and Paulev, P.-E. (1982). Muscle cell leakage due to long distance training. *European Journal of Applied Physiology*, **48**, 177–88.

Nosaka, K. and Clarkson, P.M. (1992). Relationship between post-exercise plasma CK elevation and muscle mass involved in the exercise. *International Journal of Sports Medicine*, **13**, 471–5.

Nurenberg, P., Giddings, C.J., Stray-Gundersen, J., Fleckenstein, J.L., Gonyea, W.J., and Peshock, R.M. (1992). MR Imaging-guided muscle biopsy for correlation of increased signal intensity with ultrastructural change and delayed-onset muscle soreness after exercise. *Radiology*, **184**, 865–9.

O'Donnell, M.P., Kasiske, B.L., Kim, Y., Atluru, D., and Keane, W.F. (1993). Lovastatin inhibits proliferation of rat mesangial cells. *Journal of Clinical Investigation*, **91**, 83–7.

Ogilvie, R.W., Armstrong, R.B., Baird, K.E., and Bottoms, C.L. (1988). Lesions in rat soleus muscle following eccentrically biased exercise. *American Journal of Anatomy*, **182**, 335–46.

Osterman, P.O., Askmark, H., and Wistrand, P.J. (1985). Serum carbonic anhydrase III in neuromuscular disorders and in healthy persons after a long distance run. *Journal of the Neurological Sciences*, **70**, 347–57.

Reijneveld, J.C., Ferrington, D.A., Amelink, G.J., and Bär, P.R. (1994). Estradiol treatment reduces structural muscle damage after exercise in male rats. *Medicine and Science in Sports and Exercise* **25**, S72.

Reijneveld, J.C., Koot, R.W., Bredman, J.J., Joles, J.A., and Bär, P.R. (1996). Young rats develop myopathy at lower doses of HMG-CoA reductase inhibitors than adult animals. *Pediatric Research*, **39**, 1–8.

Rodenburg, J.B., De Boer, R.W., and Bär, P.R. (1992a). Muscle damage: a combined MRI and MRS study. In *Basic and applied myology: perspectives for the 90's* (ed. U. Carraro, and S. Salmons), pp. 155–62. Unipress, Padua.

Rodenburg, J.B., Jongen, E.L.M.N., Markus, C.A.M., van der Meulen, M.F.G., Amelink, G.J., Schamp, S.M.A.A. *et al.* (1992b). Release of creatine kinase and myoglobin after eccentric exercise. Effect of exercise duration and muscle mass. *Pflügers Archiv—European Journal of Physiology*, **421**, R41.

Rodenburg, J.B., Schiereck, P., and Bär, P.R. (1992c). Effect of warming up, stretching and massage on muscle damage due to negative work. *Pflügers*

Archiv—European Journal of Physiology, **420**, R137.

Rodenburg, J.B., Bär, P.R., and De Boer, R.W. (1993). Relation between muscle soreness and biochemical and functional outcomes of eccentric exercise. *Journal of Applied Physiology*, **74**, 2976–83.

Rodenburg, J.B., Steenbeek, D., Schiereck, P., and Bär, P.R. (1994*a*). Warm-up, stretching and massage diminish harmful effects of eccentric exercise. *International Journal of Sports Medicine*, **15**, 414–19.

Rodenburg, J.B., de Boer, R.W., Schiereck, P., van Echteld, C.J., and Bär, P.R. (1994*b*). Changes in phosphorus compounds and water content in skeletal muscle due to eccentric exercise. *European Journal of Applied Physiology*, **68**, 205–13.

Rodenburg, J.B., de Boer, R.W., Jeneson, J.A., van Echteld, C.J., and Bär, P.R. (1994*c*). ^{31}P-MRS and simultaneous quantification of dynamic human quadriceps exercise in a whole body MR scanner. *Journal of Applied Physiology*, **77**, 1021–29.

Rodenburg, J.B., de Groot, M.C., van Echteld, C.J., Jongsma, HJ., and Bär, P.R. (1995). Phosphate metabolism of prior eccentrically loaded vastus medialis muscle during exercise in humans. *Acta Physiologica Scandinavica*, **153**, 97–108.

Rogers, M.A., Stull, M.A., and Apple, F.S. (1985). Creatine kinase isoenzyme activities in men and women following a marathon race. *Medicine and Science in Sports and Exercise*, **17**, 679–82.

Russell, B., Dix, D.J., Haller, D.L., and Jacobs-El, J. (1992). Repair of injured skeletal muscle. A molecular approach. *Medicine and Science in Sports and Exercise*, **24**, 189–96.

Salminen, A., Kainulainen, H., Arstila, A.U., and Vihko, V. (1984). Vitamin E deficiency and the susceptibility to lipid peroxidation of mouse cardiac and skeletal muscles. *Acta Physiologica Scandinavica*, **122**, 565–70.

Schalke, B.B., Schmidt, B., Toyka, K., and Hartung, H. (1992). Pravastatin-associated inflammatory myopathy. *New England Journal of Medicine*, **327**, 649–50.

Schultz, E. (1989). Satellite cell behavior during skeletal muscle growth and regeneration. *Medicine and Science in Sports and Exercise*, **21**, S181–6.

Serajuddin, A.T.M., Ranadive, S.A., and Mahoney, E.M. (1991). Relative lipophilicities, solubilities and structure-pharmacological considerations of HMG-CoA reductase inhibitors pravastatin, lovastatin, mevastatin and simvastatin. *Journal of Pharmaceutical Sciences*, **80**, 830–4.

Shellock, F.G., Fukunaga, T., Mink, J.H., and Edgerton, V.R (1991). Exertional muscle injury. Evaluation of concentric versus eccentric actions with serial MR imaging. *Radiology*, **179**, 659–64.

Shimomura, Y., Suzuki, M., Sugiyama, S., Hanaki, Y., and Ozaa, T. (1991). Protective effect of coenzyme Q_{10} on exercise-induced muscle damage. *Biochemical and Biophysical Research Communications*, **176**, 349–55.

Shumate, J.B., Brooke, M.H., Carroll, J.E., and Davis, J.E. (1979). Increased serum creatine kinase after exercise: a sex linked phenomenon. *Neurology*, **29**, 902–4.

Sjöström, M. and Fridén, J. (1984). Muscle soreness and muscle structure. *Medicine and Science in Sports and Exercise*, **17**, 169–86.

Smit, J.W., Bär, P.R., Geerdink, R.A., and Erkelens, D.W. (1995). Heterozygous familial hypercholesterolaemia is associated with pathological exercise-induced leakage of muscle proteins, which is not aggravated by simvastatin therapy. *European Journal of Clinical Investigation*, **25**, 79–84.

Smith, L.L. (1991). Acute inflammation; the underlying mechanism in delayed onset muscle soreness? *Medicine and Science in Sports and Exercise*, **23**, 542–51.

Smith, P.F., Eydelloth, R.S., Grossman, S.J., Stubbs, R.J., Schwartz, M.S., Germershausen, J.I. *et al.* (1991). HMG-CoA reductase inhibitor-induced myopathy in the rat. Cyclosporin A interaction and mechanism studies. *Journal of Pharmacology and Experimental Therapeutics*, **257**, 1225–35.

Soza, M., Karpati, G., Carpenter, S., and Prescott, S. (1986). Calcium-induced damage of skeletal muscle fibres is markedly reduced by calcium channel blockers. *Acta Neuropathologica*, **71**, 70–5.

Stauber, W.T., Clarkson, P.M., Fritz, V.K., and Evans, W. (1990). Extracellular matrix disruption and pain after eccentric action. *Journal of Applied Physiology*, **69**, 868–74.

Sugie, H. and Verity, M.A. (1985). Postnatal histochemical fiber type differentiation in normal and hypothyroid rat soleus muscle. *Muscle and Nerve*, **8**, 654–60.

Takala, T., Rahkila, P., Hakala, E., Vuori, J., Puranen, J., and Väänänen, K. (1989). Serum carbonic anhydrase III, an enzyme of type 1 muscle fibres, and the intensity of physical exercise. *Pflügers Archiv—European Journal of Physiology*, **413**, 447–50.

Tiidus, P.M. (1995). Can estrogens diminish exercise induced muscle damage? *Canadian Journal of Applied Physiology*, **20**, 26–38.

Tiidus, P.M. and Houston, M.E. (1993). Vitamin E status does not affect the responses to exercise training and acute exercise in female rats. *Journal of Nutrition*, **123**, 834–40.

Väänänen, H.K., Leppilampi, M., Vuori, J., and Takala, T.E.S. (1986). Liberation of muscle carbonic anhydrase into serum during extensive exercise. *Journal of Applied Physiology*, **61**, 561–4.

Vater, R., Cullen, M.J., and Harris, J.B. (1992). The fate of desmin and titin during the degeneration and regeneration of the soleus muscle of the rat. *Acta Neuropathologica*, **84**, 278–88.

Volfinger, L., Lassourd, V., Michaux, J.M., Braun, J.P., and Toutain, P.L. (1994). Kinetic evaluation of muscle damage during exercise by calculation of amount of creatine kinase released. *American Journal of Physiology*, **266**, R434–41.

Warren, J., Jenkins, R.R., Packer, L., Wit, E.H., and Armstrong, R.B. (1992). Elevated muscle vitamin E does not attenuate eccentric exercise-induced muscle injury. *Journal of Applied Physiology*, **72**, 2168–75.

Wenos, J.Z., Brilla, L.R., and Morrison, M.J. (1990). Effect of massage on delayed onset muscle soreness. *Medicine and Science in Sports and Exercise*, **22**, S34.

Wernig, A., Salvini, T.F., and Irintchev, A. (1991). Axonal sprouting and changes in fibre types after running-induced muscle damage. *Journal of Neurocytology*, **20**, 903–13.

White, T.P. and Esser, K.A. (1989). Satellite cell and growth factor involvement in skeletal muscle growth. *Medicine and Science in Sports and Exercise*, **21**, S158–63.

Wiseman, H. (1994). Tamoxifen: new membrane-mediated mechanisms of action and therapeutic advances. *Trends in Pharmacological Sciences*, **15**, 83–9.

Wrogeman, K. and Pena, S.D.J. (1976). Mitochondrial calcium overload: a general mechanism for cell-necrosis in muscle diseases. *Lancet*, **i**, 672–3.

Yamada, S., Buffinger, N., Domario, J., and Strohman, R.C. (1989). Fibroblast growth factor is stored in fiber extracellular matrix and plays a role in regulating muscle hypertrophy. *Medicine and Science in Sports and Exercise*, **21**, 173–80.

2

Muscle damage induced by contraction: an in situ single skeletal muscle model

John A. Faulkner and Susan V. Brooks

2.1   Introduction

At the turn of the century, during an investigation of fatigue of the finger flexors, Hough (1901) described the phenomenon of late onset muscle soreness. The next year, in a study of muscle soreness, he attributed the phenomenon to injury and necrosis of muscle fibres (Hough 1902). Eighty years later, direct evidence of cellular and ultrastructural damage was demonstrated in the skeletal muscles of human beings (Fridén *et al.* 1983; Newham *et al.* 1983a), mice (Wernig *et al.* 1990), and rats (Ogilvie *et al.* 1988) after pliometric exercise, that is, exercise during which most of the physical activity involves stretching of the contracting muscles (for example, running downhill). Evidence for contraction-induced injury to skeletal muscles also comes from a variety of indirect measures, including creatine kinase release, calcium influx, activation of the lysosomal system, increased glutathione disulfide, and an impaired capacity to develop force (Armstrong 1984). The time courses of the direct and indirect measures of injury and, in the case of human subjects, of reported muscle soreness indicate an increase in the severity of the injury several days after the actual occurrence.

These data support Hough's (1902) premise of muscle fibre necrosis in muscles that have not been trained for pliometric exercise. Although they provide evidence of injury, the studies of pliometric exercise provide no definitive information about the recruitment patterns within muscle groups, or the magnitude, direction, or velocity of the displacements to which muscle fibres must be exposed to initiate a given injury. An *in situ* extensor digitorum longus (EDL) muscle preparation was therefore developed to obtain more precise data about the circumstances under which contractions injure muscle fibres and to test rigorously hypotheses as to the underlying mechanisms (McCully and Faulkner 1985). Throughout this chapter, 'contraction' refers to maximum activation of a muscle, producing a tendency to shorten. Whether the muscle actually shortens, remains isometric, or is lengthened depends on the load. For brevity, these conditions will be referred to as miometric, isometric, and pliometric contractions, respectively.

2.2 Methods

The single muscle preparation is shown in Fig. 2.1. The knee of the anaesthetized mouse is pinned to a baseplate, and the tendon is attached to the lever arm of a servomotor-force transducer. The EDL muscle, which contains mainly fast fibres, is activated by stimulation of the peroneal nerve. This arrangement provides control of the magnitude and rate of displacement of the muscle, and measurement of the force developed (Fig. 2.2).

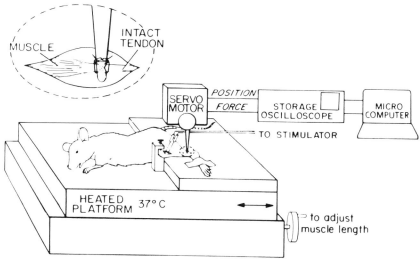

Fig. 2.1 *In situ* muscle preparation for isometric, miometric, and pliometric contractions and passive lengthening of extensor digitorum longus muscles from mice. Mice were anaesthetized with pentobarbital sodium while in the apparatus. (Reprinted from Faulkner *et al.* 1990, with permission.)

An initial hypothesis was that contraction-induced injury could occur during any type of contraction, but that injury was most likely to occur, and would be most severe, during stretch of a maximally activated muscle. To test this hypothesis, McCully and Faulkner (1985) developed comparable protocols of miometric, isometric, and pliometric contractions (Fig. 2.2). EDL muscles were either shortened or lengthened through a distance equivalent to 20 per cent of fibre length (L_f) at a rate of 0.5 to 1 L_f/s. They were stimulated for 300 to 500 ms at a frequency that resulted in approximately 85 per cent of the maximum tetanic force (P_o) developed when contraction took place under isometric conditions. Contractions were elicited every 4 to 5 s over periods from 5 min (Zerba *et al.* 1990) to 30 min (Faulkner *et al.* 1989), or for 5 min with 5 min of rest, repeated three times (McCully and Faulkner 1985; Brooks and Faulkner 1990). In different studies the number of pliometric contractions ranged from 75 to 360. After exposure to a contraction protocol, or to passive lengthening, the incisions were closed and the mice allowed to recover. A specified time was allowed to elapse before the animals were again anaesthetized and the same apparatus was used to evaluate the capacity of the muscles to develop force. Muscles were then removed and sections fixed for evaluation of the damage by light microscopy (Zerba *et al.* 1990) or electron microscopy (Brooks *et al.* 1995). The mice were killed by anaesthetic overdose.

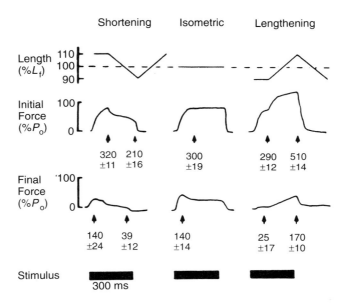

Fig. 2.2 Representative recordings of length (upper traces) and force (lower traces) for miometric, isometric, and pliometric contractions. For all traces, muscles were stimulated for 300 ms at 150 Hz. Recordings from the beginning of the stimulation protocol are shown in upper force traces and recordings from the end of the stimulation protocol are shown in lower force traces. Mean ± SEM of force in mN is indicated under force traces. Force measurements shown are the peak forces for the isometric portion of all contractions and the forces measured at the end of the miometric and pliometric contractions. (Modified from McCully and Faulkner 1985, with permission.)

2.3   Criteria of injury

Direct measurement of muscle injury has to be based on morphological evidence of fibre damage (Fig. 2.3). An effective, but indirect, measure of damage is the decrease in the value of P_o developed by muscles at selected times after a protocol of contractions. The value of P_o after the protocol is normalized by expressing it as a percentage of the initial control value

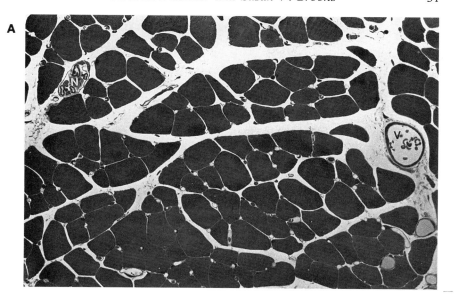

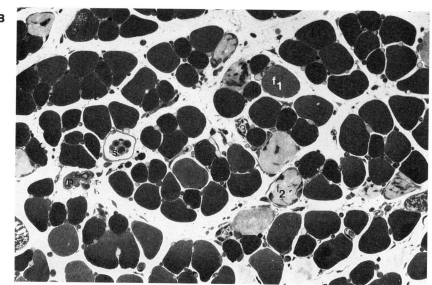

Fig. 2.3 Cross-sections through the belly of (A) an uninjured and (B) an injured extensor digitorum longus muscle stained with toluidine blue at 3 days post-injury. In uninjured muscle, fibres were angular in shape and showed a uniform intensity of staining. In injured muscle, evidence of damage ranged from minor changes in staining intensity and rounded appearance of fibres (f_1) to various stages of phagocytotic infiltration (f_2). Some fibres contained phagocytes inside the old basal laminae. Of these fibres some were completely devoid of cytoplasmic organelles (f_3). s, Muscle spindle; n, nerve; v, blood vessel. Magnification, 360 ×. (Reproduced from Zerba *et al.* 1990, with permission.)

(Fig. 2.4). The 'force deficit' observed at any time after the protocol is calculated as the difference between the observed value for P_o and the control value, expressed as a percentage of the control value. Three days after a pliometric contraction protocol, the magnitude of the force deficit correlates ($r = 0.7$) with the number of, or area occupied by, severely damaged fibres (McCully and Faulkner 1986; Zerba *et al.* 1990). Examples of damaged fibres are given in Fig. 2.3. In spite of the relationship between the number of damaged fibres and the force deficit, the force deficit at day 3 is about 15 per cent greater than the extent of the damage observed in histological sections (McCully and Faulkner 1986; Zerba *et al.* 1990). This discrepancy is most probably due to the focal nature of the damage within single fibres (Stauber *et al.* 1988), and to ultrastructural damage that impairs force development but is not observable with light microscopy. We conclude that the force deficit provides the better estimate of the totality of contraction-induced injury.

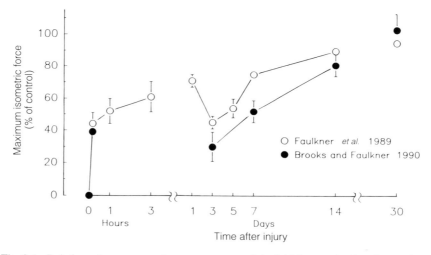

Fig. 2.4 Relative values, expressed as a percentage of the initial control value, for maximum isometric tetanic force (P_o) developed by extensor digitorum longus (EDL) muscles from young mice (3 months old) at selected time periods following a protocol of pliometric contractions known to produce injury to muscle fibres. The values indicated on the abscissa are the times in hours and days after the final contraction. The value at time zero is the relative P_o developed prior to the initiation of the last pliometric contraction. The open circles represent values obtained following a pliometric contraction protocol administered during plantar flexion of the ankle joint with the foot in a 'shoe' apparatus, with contractions every 5 s for 30 min. The closed circles represent values obtained with an *in situ* muscle preparation with the intact tendon of the EDL muscle attached to the lever arm of a servomotor and the contracting muscle stretched through 20 per cent of fibre length. The protocol consisted of a pliometric contraction every 4 s for 3 bouts of 5 min each with 5 min rest between bouts. All values are expressed as mean ± SEM. (Redrawn from data in Brooks and Faulkner 1990 and Faulkner *et al.* 1989.)

2.4   Sequence of events

The sequence of events that follows protocols consisting of repeated miometric, isometric, or pliometric contractions, or of repeated short stretches of
passive muscles, differs greatly. The repeated short stretches of passive
muscles produce neither fatigue nor injury (McCully and Faulkner 1985;
Faulkner *et al.* 1989). Immediately after the completion of comparable
protocols, the loss in force is greatest for pliometric, less for miometric,
and least for isometric contractions (Fig. 2.2). Miometric and isometric
contractions produce significant fatigue immediately and for several hours
after the protocols, but no evidence of injury at day 3 (McCully and Faulkner
1985; Faulkner *et al.* 1989). In contrast, after pliometric contraction protocols, muscles show morphological evidence of injury throughout the first 5
days. The magnitude of the injury is a function of the duration of the
pliometric exercise, or the number of pliometric contractions (Brooks and
Faulkner 1990). In histological sections, the greatest damage is observed at
day 3 (McCully and Faulkner 1985; Faulkner *et al.* 1989; Zerba *et al.* 1990).
Impaired force development is observed for up to 30 days after the initial
injury (Fig. 2.4). These data support the hypothesis that injury is more likely
to occur during pliometric contractions, and is more severe after pliometric
contractions than after miometric or isometric contractions.

Immediately after a protocol of 75 pliometric contractions, P_o decreases to
zero, although force may still be developed during stretches of passive
muscles. The inability of the muscle to develop isometric force is the result
of both fatigue and the initial injury. The recovery of force development
during the first few hours is rapid, but then reaches a plateau between 3 and
24 h. The increase in P_o to a stable value marks the recovery from fatigue,
which is complete within 24 hours, at most (Faulkner and Brooks 1993). The
average of the force deficits measured at 3 and 24 h provides the best estimate
of the magnitude of the initial injury. This calculation gives an initial force
deficit of 35 per cent immediately after a protocol of pliometric contractions
that results in a maximum force deficit of 55 per cent on day 3 (Fig. 2.4).
After the initial injury, a secondary (and more severe) injury is observed
between day 1 and day 3. After day 3, the recovery of force is gradual and is
usually complete within 30 days (Fig. 2.4).

2.4.1   Initial injury

The initial injury produced by stretching a muscle appears to result from
mechanical damage to the ultrastructure of specific sarcomeres. The repeated
contraction protocols do not permit rigorous testing of hypotheses regarding
the relationship between mechanical factors and the initiation of injury. The
requirements for such a test are:

(1) a close association in time between the two events;
(2) clear identification of specific mechanical events within the contracting muscle;
(3) an accurate and immediate measure of the totality of the injury.

The only possibility for an immediate measure of the totality of the injury was a measurement of force deficit, free from the complications of fatigue. A single stretch would not produce fatigue, and we concluded that this would provide the clearest identification of the average and peak force developed and the magnitude of the strain to which the muscle had been subjected. We therefore developed a protocol in which EDL muscles of mice were subjected *in situ* to single stretches under passive or maximally activated conditions (Brooks *et al.* 1995). Injury was analysed one minute after single stretches by evaluating the force deficit and by examining the ultrastructural damage to sarcomeres. With single stretches, passive muscles displayed a force deficit only when exposed to strains beyond 50 per cent of fibre length, which is beyond the physiological range. In contrast, maximally activated muscles showed force deficits at strains of 30 per cent and more. From these data, multiple linear regression equation models were used to estimate the relationships of strain, peak force, average force, and work, on the one hand, to the size of the force deficit, on the other. When interactions among variables were ignored, each variable had a significant relationship with the force deficit that followed the single stretch (Brooks *et al.* 1995). The relative importance of the independent variables on the force deficit was established with a stepwise regression analysis, which indicated that the work done to stretch the muscle accounted for the largest part of the variation in the force deficit, explaining 85 and 76 per cent of the variation in the force deficit for passive and maximally activated muscles, respectively (Fig. 2.5).

Although work was a strong predictor of damage, the relationship between the force deficit and work changed with activation (Fig. 2.5). Under passive conditions, muscles were injured by large stretches, but this happened at forces that were often less than P_o. Maximally activated muscles were injured by relatively small stretches, but at average forces that were between two- and threefold greater than P_o. For active muscles, the high average forces appear to influence the distribution of muscle strain among different sarcomeres in series along the length of myofibrils. During pliometric contractions, sarcomere lengths become more heterogeneous than they are at rest or during shortening or isometric contractions (Burton *et al.* 1989). Our working hypothesis is that, within activated muscle fibres, weak and strong sarcomeres exist in series. An increase in the differences between the strengths of sarcomeres in series increases the likelihood of the weaker sarcomeres being subjected to large strains and subsequent injury even when the activated muscle fibre undergoes a stretch of relatively small magnitude.

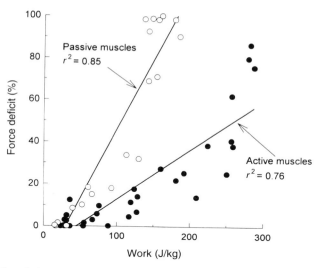

Fig. 2.5 The relationship between the force deficit observed 1 minute after single stretches of passive (open circles) and maximally activated (filled circles) muscles and the work done to lengthen the muscle. Data are for single stretches varying in magnitude but not velocity ($V = 2L_f/$ s). Each symbol indicates a data point from a single stretch. The r^2 values represent the coefficients of determination for the regression relationships. The slopes of the relationships, 0.23 for active muscles and 0.66 for passive muscles, are significantly different. Sample size is 20 for passive muscles and 32 for active muscles. (Reprinted from Brooks *et al.* 1995.)

Morgan (1990) has described the phenomenon of the stretching of weak sarcomeres beyond the point of overlap of thick and thin filaments as 'popping'. Several investigators have suggested that injury is initiated when the actin and myosin filaments of individual sarcomeres are 'pulled apart' (Newham *et al.* 1983b; Higuchi *et al.* 1988; Wood *et al.* 1993). Furthermore, Higuchi *et al.* (1988) demonstrated that repeated stretches of passive single fibres result in some sarcomeres being stretched beyond overlap, with resultant injury to the overstretched sarcomeres. With the assumption of thick and thin filament lengths of 1.6 μm and 1.02 μm, respectively, overlap of thick and thin filaments would cease to exist at a sarcomere length of 3.64 μm. This corresponds to a sarcomere strain of about 50 per cent, expressed relative to the average sarcomere length of 2.52 μm reported for EDL muscles of mice at the physiological resting length (Brooks and Faulkner 1994). Support for the hypothesis that muscle fibres are injured when sarcomeres are stretched beyond the point of filament overlap is provided by the observation that, for passive muscles, single stretches of 50 per cent strain are necessary to produce a significant force deficit, and stretches of 60 per cent strain result in almost complete elimination of force production (Fig. 2.5). Once sarcomeres are stretched beyond filament overlap, there are no differences between stretched sarcomeres in active and passive fibres. Consequently, the mechanism underlying the injury to individual sarcomeres, and the dependence of the extent of the injury on the magnitude of the stretch beyond overlap of thick and thin filaments, would not be different in the two cases. Ultrastructural injury to small groups of sarcomeres following single (Fig. 2.6) or repeated (Fridén *et al.* 1983; Newham *et al.* 1983b) pliometric contractions provides indirect support for the hypothesis that injury occurs to small groups of weak sarcomeres, but the hypothesis has not been tested directly.

2.4.2 Secondary injury

Evidence for the existence, timing, and magnitude of the secondary injury comes from the decrease between day 1 and day 3 in the capacity of muscles for developing force (Fig. 2.4). The extent of the fibre damage observed in histological sections (McCully and Faulkner 1985; Ogilvie *et al.* 1988; Zerba *et al.* 1990) provides further support for the secondary injury occurring at this time. Although the evidence is more qualitative, ultrastructural damage (Fridén *et al.* 1983; Newham *et al.* 1983b) and reported delayed onset muscle soreness in human subjects (Newham *et al.* 1983a, b) are also most severe 2 to 3 days after pliometric exercise. The severity of the injury to the ultrastructural elements of different fibres varies from thickening of the Z-disc, observed in human muscle after pliometric exercise (Burton *et al.* 1989), to complete elimination of all sarcoplasmic elements in our own experiments (Fig. 2.3). The secondary injury is associated with a cascade of events that includes an inflammatory response (Fig. 2.3), further degeneration of portions of fibres, and subsequent phagocytosis of the myofilaments and sarcoplasmic elements (Zerba *et al.* 1990). An increase in the concentration of creatine kinase in the plasma (Newham *et al.* 1983a) further suggests that the integrity of the plasmalemma and basement membrane is disrupted.

The magnitude of the force deficit of 45 per cent associated with the secondary injury is reduced to 15 per cent when young mice are treated with the oxygen free-radical scavenger polyethylene glycol superoxide dismutase (PEG-SOD) before the EDL muscle is exposed to the pliometric contraction protocol and during the 3 subsequent days (Zerba *et al.* 1990). This supports the hypothesis that the secondary injury results primarily from oxygen free radicals that aggravate the damage to fibres already affected by the mechanical injury. The magnitude of the secondary injury and the significant protection provided by PEG-SOD suggests that oxygen free radicals cause damage that exceeds the degradation of previously damaged portions of fibres.

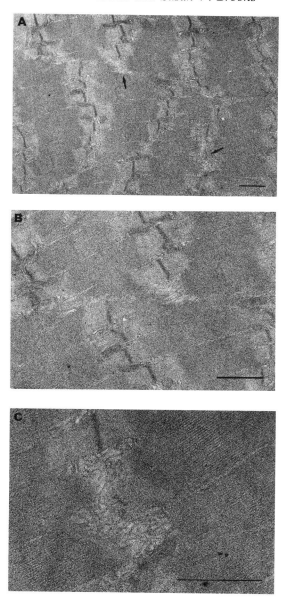

Fig. 2.6 Longitudinal section of a fibre immediately after a single 60 per cent stretch of a maximally activated EDL muscle. The injured fibre shows focal abnormalities, with misaligned and overstretched sarcomeres, disruption or absence of Z-disc material, thin filaments displaced from thick filaments, and disrupted striation patterns. Sarcomeres in series range from severely damaged to normal. Panels B and C show, at higher levels of magnification, the two regions in panel A that are indicated with arrows. For panel A, the magnification is 7128 ×; for panel B, 12 420 ×; and for panel C, 22 680 ×. All scale bars represent 1 µm. (Reprinted from Brooks *et al.* 1995.)

2.4.3   Recovery from injury

Recovery from the injury begins at about day 3 and progresses to a complete return to normal structure and function. Complete recovery occurs between day 7 and day 30, depending on the severity of the secondary injury (McCully and Faulkner 1985; Faulkner *et al.* 1989; Brooks and Faulkner 1990). The recovery of structure and function of damaged fibres appears to be the result of regeneration (Carlson and Faulkner 1983). This process involves the activation and subsequent division of satellite cells to form myoblasts. The myoblasts fuse to form myotubes which mature into undifferentiated myofibres. If they are innervated, the myofibres differentiate into specific fibre types (Carlson and Faulkner 1983). After contraction-induced injury, activated satellite cells have been observed, particularly near the myotendinous junctions of growing and mature skeletal muscle fibres (Darr and Schultz 1987). Myotubes were first noted in regenerating fibres 4 to 5 days after a pliometric contraction protocol that injured 35 per cent of the muscle fibres in a given section (McCully and Faulkner 1985).

Regeneration of fibres occurs within the surviving basement membrane (Caldwell *et al.* 1990). Whether minor injuries to Z-discs, or to only a few sarcomeres within a fibre, are repaired by the same process of satellite cell activation is not known.

Acknowledgements

We thank Gabriele Wienert and Richard Hinkle for their assistance in the preparation of this chapter. The research upon which the review is based was supported in part by the National Institute on Aging RO1 grant AG-06157 and by a Multidisciplinary Training Grant on Aging AG-00114 that provided a traineeship to S.V.B.

References

Armstrong, R.B. (1984). Mechanisms of exercise-induced delayed onset muscular soreness: a brief review. Medicine and Science in Sports and Exercise, **16**, 529–38.

Brooks, S.V. and Faulkner, J.A. (1990). Contraction-induced injury: recovery of skeletal muscles in young and old mice. *American Journal of Physiology*, **258**, C436–42.

Brooks, S.V. and Faulkner, J.A. (1994). Isometric, shortening and lengthening contractions of muscle fiber segments from adult and old mice. *American Journal of Physiology*, **267**, C507–13.

Brooks, S.V., Zerba, E., and Faulkner, J.A. (1995). Injury to fibres after single stretches of passive and maximally stimulated muscles in mice. *Journal of Physiology (London)*, **488**, 459–69.

Burton, K., Zagotta, W.N., and Baskin, R.J. (1989). Sarcomere length behaviour along single frog muscle fibres at different lengths during isometric tetani. *Journal of Muscle Research and Cell Motility*, **10**, 67–84.

Caldwell, C.J., Mattey, D.L., and Weller, R.O. (1990). Role of the basement membrane in the regeneration of skeletal muscle. *Neuropathology and Applied Neurobiology*, **16**, 225–38.

Carlson, B.M. and Faulkner, J.A. (1983). The regeneration of skeletal muscle fibers following injury: a review. *Medicine and Science in Sports and Exercise*, **15**, 187–98.

Darr, K.C. and Schultz, E. (1987). Exercise-induced satellite cell activation in growing and mature skeletal muscle. *Journal of Applied Physiology*, **63**, 1816–21.

Faulkner, J.A. and Brooks, S.V. (1993). Fatigability of mouse muscles during constant length, shortening, and lengthening contractions: interactions between fiber types and duty cycles. In *Neuromuscular fatigue* (ed. A.J. Sargeant and D. Kernell), pp. 116–23. Elsevier, Amsterdam.

Faulkner, J.A., Jones, D.A., and Round, J.M. (1989). Injury to skeletal muscles of mice by forced lengthening during contractions. *Quarterly Journal of Experimental Physiology*, **74**, 661–70.

Faulkner, J.A., Zerba, E., and Brooks, S.V. (1990). Contraction induced injury to skeletal muscle fibers. In *Hypoxia: the adaptations* (ed. J.R. Sutton, G. Coates, and J.E. Remmers), pp. 225–30. B.C. Decker, Toronto.

Fridén, J., Sjöström, M., and Ekblom, B. (1983). Myofibrillar damage following intense eccentric exercise in man. *International Journal of Sports Medicine*, **4**, 170–76.

Higuchi, H., Yoshioka, T., and Maruyama, K. (1988). Positioning of actin filaments and tension generation in skinned muscle fibres released after stretch beyond overlap of the actin and myosin filaments. *Journal of Muscle Research and Cell Motility*, **9**, 491–8.

Hough, T. (1901). Ergographic studies in neuro-muscular fatigue. *American Journal of Physiology*, **5**, 240–66.

Hough, T. (1902). Ergographic studies in muscular soreness. *American Journal of Physiology*, **7**, 76–92.

McCully, K.K. and Faulkner, J.A. (1985). Injury to skeletal muscle fibers of mice following pliometric contractions. Journal of Applied Physiology, **59**, 119–26.

McCully, K.K. and Faulkner, J.A. (1986). Characteristics of pliometric contractions associated with injury to skeletal muscle fibers. *Journal of Applied Physiology*, **61**, 293–99.

Morgan, D.L. (1990). New insights into the behavior of muscle during active lengthening. *Biophysical Journal*, **57**, 209–21.

Newham, D.J., Jones, D.A., and Edwards, R.H.T (1983a). Large delayed plasma creatine kinase changes after stepping exercise. *Muscle and Nerve*, **6**, 380–5.

Newham, D.J., McPhail, G., Mills, K.R., and Edwards, R.H.T. (1983b). Ultrastructural changes after concentric and eccentric contractions of human muscle. *Journal of the Neurological Sciences*, **61**, 109–22.

Ogilvie, R.W., Armstrong, R.B., Baird, K.E., and Bottoms, C.L. (1988). Lesions in the rat soleus muscle following eccentrically biased exercise. *American Journal of Anatomy*, **182**, 335–46.

Stauber, W.T., Fritz, V.K., Vogelbach, D.W., and Dahlmann, B. (1988). Characterization of muscles injured by forced lengthening. I. Cellular infiltrates. *Medicine
and Science in Sports and Exercise*, **20**, 345–54.

Wernig, A., Irintchev, A., and Weisshaupt, P. (1990). Muscle injury, cross-sectional
area and fibre type distribution in mouse soleus after intermittent wheel-running.
Journal of Physiology (London), **428**, 639–52.

Wood, S.A., Morgan, D.L., and Proske, U. (1993). Effects of repeated eccentric
contractions on structure and mechanical properties of toad sartorius muscle.
American Journal of Physiology, **265**, C792–800.

Zerba, E., Komorowski, T.E., and Faulkner, J.A. (1990). Free radical injury to
skeletal muscles of young, adult, and old mice. *American Journal of Physiology*,
258, C429–35.

Muscle damage induced by cyclic eccentric contraction: biomechanical and structural studies

Jan Fridén and Richard L. Lieber

3.1 Background

It is widely agreed that muscle injury, soreness, and elevated serum enzyme levels are associated with exercise that involves eccentric contractions. One of the earliest methods for inducing such muscle injury experimentally used a motor-driven exercise ergometer (Fig. 3.1), and led to the demonstration of ultrastructural abnormalities within the myofibrils (Fridén *et al.* 1981). Other authors have observed similar myofibrillar changes after repetitive eccentric contractions (Fridén *et al.* 1981, 1983, 1988; Armstrong *et al.* 1983; Kuipers *et al.* 1983; Newham *et al.* 1983). The characteristic pattern of changes in the contractile apparatus includes broadening, smearing, or even total disruption of Z-discs (Fig. 3.2); the Z-discs of adjacent myofibrils lose register and follow a zigzag course which is referred to as 'Z-disc streaming'. In addition, A-bands within affected sarcomeres may be out of register and, in some cases, thick filaments may be completely absent. The biological significance of the Z-disc streaming that occurs as a result of eccentric contraction is currently unknown. It is, however, a common finding in neuromuscular disorders (Mastaglia and Walton 1982). Z-disc streaming appears to be a primary myofibrillar response to altered physical and metabolic conditions.

Given these initial observations, we developed a model system to study the phenomenon in order to elucidate the underlying mechanisms. This review will focus on the biomechanical factors and cellular changes that result from eccentric contraction-induced muscle injury in a skeletal muscle of the rabbit. Before discussing the specific experimental observations, we will review the basic biomechanical properties of skeletal muscle.

3.2 Skeletal muscle biomechanics

Two fundamental sarcomeric properties are used to define the isometric and isotonic contractile properties of muscle: the length-tension relationship and the force-velocity relationship.

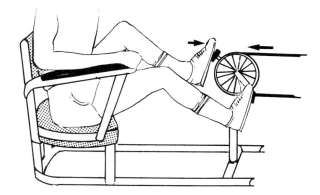

Fig. 3.1 Motor-driven bicycle ergometer used to create eccentric contraction of the human quadriceps femoris muscles. The individual pushes on the pedals (short arrow) and is resisted by the motor drive (long arrow). (From Lieber 1992.)

Fig. 3.2 Ultrastructural abnormalities in the extensor digitorum longus (EDL) muscle 2 days after eccentric contractions. The injured region includes disruption of the Z-discs (smearing and streaming) and A-bands. Calibration bar, 5 μm.

vimentin (human monoclonal antibody) to demonstrate inflammatory cells (Lazarides 1982), desmin (monoclonal mouse antihuman) for evaluation of the structural integrity of the cytoskeletal network (Thornell *et al.* 1985), and embryonic myosin (monoclonal goat-antimouse antibody) to detect the presence of regenerating fibres. Antibody binding was visualized by the indirect peroxidase-antiperoxidase technique (Sternberger 1979).

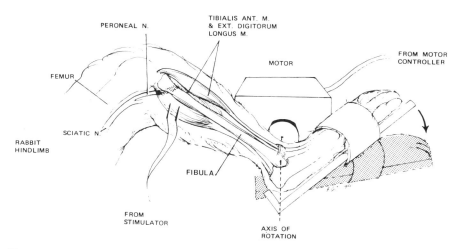

Fig. 3.5 Method for creating an eccentric contraction-induced injury non-invasively. Percutaneous electrodes are used to stimulate the peroneal nerve during application of a torque that produces ankle plantarflexion (heavy arrow).

3.3.4 Cytoskeletal changes

The most striking change, observed soon after eccentric contraction, was the loss of desmin labelling in many muscle fibres across the muscle cross-section (Fig. 3.6). At 1 and 3 days post-contraction, desmin labelling was absent from most of the fibres in some portions of the muscle cross-section while the loss of labelling was more scattered in other portions (Fig. 3.7(A)). This loss was significantly less marked 14 days after exercise and had recovered almost completely by 28 days. Fibronectin could be localized within fibres from 1 h up to 7 days post-exercise (Fig. 3.7(B)). In agreement with the data from force measurement, which demonstrated a greater force decline in the EDL than in the TA (see below), the percentage of damaged fibres within a section was much greater in the EDL than in the TA.

The loss of labelling of the structural protein desmin therefore proved to be a sensitive marker for muscle fibre damage. The structural integrity of the Z-disc and the transverse alignment of sarcomeric striations have been attributed to the existence of filamentous bridges between Z-discs and

between M-lines across the fibre axis (Price and Sanger 1979). Tokuyasu *et al.*
(1982) provided the first evidence that the connections between Z-discs were
composed of desmin. Wang and Ramirez-Mitchell (1983) showed by means
of scanning electron microscopy that the longitudinally orientated intermediate filaments originated from the periphery of Z-discs in series. In the
light of recent studies it is likely that the titin molecule plays the more
important role in limiting extreme changes of sarcomere length (Horowits
et al. 1986).

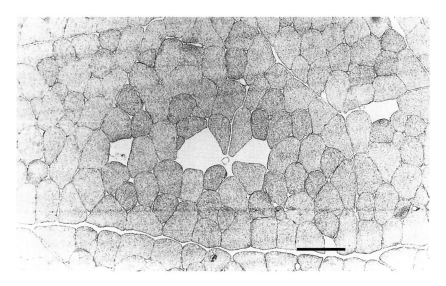

Fig. 3.6 Anti-desmin stained cryosection of injured EDL muscle 3 days after eccentric exercise.
Note the complete loss of desmin from some fibres. Calibration bar, 100 μm.

3.3.5 Time course of morphological changes

1 hour   The cross-sections contained fibres of various sizes and staining
intensities. Fibres of abnormal shape (rounded, large and lightly stained or
small and heavily stained with irregular membrane contours) were found
interspersed between the fibres of uniform size. These abnormally shaped
fibres demonstrated partial or total loss of desmin. From serial sections
stained with antibody against laminin it could be demonstrated that these
fibres were surrounded by either intact or partially interrupted basal laminae
and were positively stained with fibronectin.

1 day   An increased number of non-muscle cells was observed both around
and within the muscle fibres. Fibre size was highly variable. Laminin and

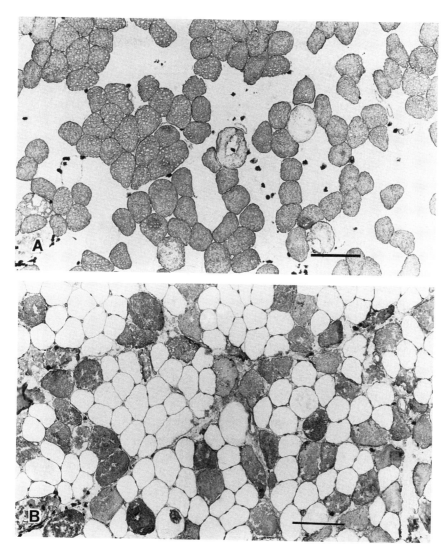

Fig. 3.7 Serial sections of injured TA muscle 1 day after eccentric exercise. (A) Anti-desmin staining: note injured fibres in which desmin stain is lost, either partially or completely. (B) Anti-fibronectin staining: note fibres demonstrating intracellular localization of plasma fibronectin, indicative of sarcolemmal disruption. Calibration bar, 100 µm. (From Lieber and Fridén 1993.)

fibronectin labelling of 'desmin-negative' fibres (that is, fibres that did not stain for desmin) showed partial or complete loss of fibre boundaries, with clusters of stain at the site of the lesion. Endomyseal and perimyseal spaces were wider than normal. Staining the desmin-negative fibres for ATPase and

NADH demonstrated a nonhomogeneous distribution of activity, with focal deposits or absence of stain within the fibres. Typing of the desmin-negative fibres could not be performed readily because of this structural damage.

3 days   A large population of fibres with extreme sizes and abnormal shapes was observed (Fig. 3.8). In many cases these fibres had been invaded by inflammatory cells. The abnormal fibres consistently stained heavily with ATPase at pH 10.3, suggesting that they were of the fast type. Based on the large proportion of very small muscle fibres, this appears to be the post-exercise time point when most of the degeneration and regeneration is taking place. A stereotypic topographical distribution of injury was seen: most injury occurred in the superficial muscle region, whereas the deeper regions were relatively spared. Of the fibres that stained strongly with the antibody to embryonic myosin there were more in the periphery of fascicles than deep within fascicles. Fibres that were positive for embryonic myosin were sometimes found in the central portions of the fascicles in the affected areas, but in a scattered manner.

7 days   A substantial fraction of the fibre population consisted of small fibres (defined as fibres that had an area less than 50 per cent of the average for control muscle fibres). Although most small fibres expressed embryonic myosin (Fig. 3.9), this was not always the case. Small fibres were, however, stained intermediately dark to dark with the anti-desmin antibody. Not only were the embryonic fibres comparatively small: they were also extremely variable in shape and consistently different to the polygonal appearance of normal adult cells (Fig. 3.10). Some embryonic myosin-positive fibres were strongly vimentin-positive; others were vimentin-negative.

14 days   The number of embryonic myosin-positive fibres was still large, although smaller than at 7 days. These fibres demonstrated normal labelling for desmin and stained more weakly for embryonic myosin than those at 7 days post-contraction. Type 1 fibres never contained embryonic myosin. All fibres expressing embryonic myosin were of type 2 but could not be categorized neatly as type 2A or 2B; they could be defined as 2AB because they were intermediately stained at pH 4.6 and 4.3.

28 days   Muscle fibres were densely packed and polygonal in shape. Small fibres were still present, and very small fibres were darkly or intermediately stained with embryonic myosin. Desmin labelling was of essentially normal intensity.

3.3.6   Torque changes

The structural changes described above were associated with functional changes. Torque in the direction of dorsiflexion decreased to a minimum

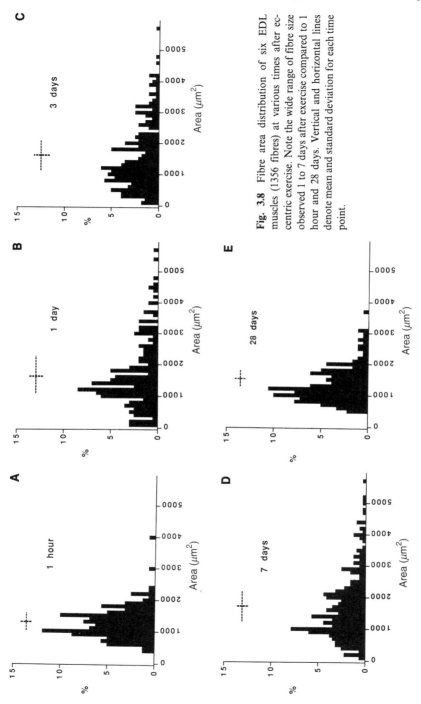

Fig. 3.8 Fibre area distribution of six EDL muscles (1356 fibres) at various times after eccentric exercise. Note the wide range of fibre size observed 1 to 7 days after exercise compared to 1 hour and 28 days. Vertical and horizontal lines denote mean and standard deviation for each time point.

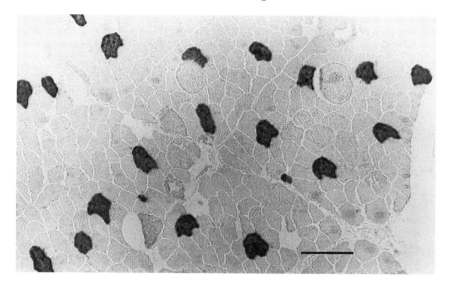

Fig. 3.9 Section of injured EDL 7 days after exercise, stained with anti-embryonic myosin antibody. Note the darkly stained embryonic myosin-positive fibres together with other physically damaged but unstained fibres. Calibration bar, 100 μm.

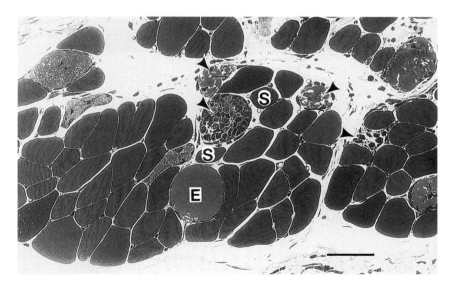

Fig 3.10 Light micrograph of biopsy taken 3 days after eccentric contractions, showing abnormally enlarged (E) and very small (S) surrounding fibres. This micrograph helps to explain the abnormal fibre size distribution in the histograms of Fig. 3.8. Several fibres are infiltrated with inflammatory cells (arrowheads). Calibration bar, 100 μm.

about 2–3 days after eccentric contraction and took about 7 days to recover completely (Fig. 3.11(A)). Although isometric contractions performed at the same intensity resulted in the identical torque decrease after 1 day, torque had recovered almost completely 2 days later. These data suggest that the decline in torque 1 day after eccentric contraction is simply the result of intramuscular substrate depletion. The amplitude of the rectified, integrated electromyogram was unchanged, so fatigue of transmission at the neuro-muscular junction could be ruled out as a potential explanation of force

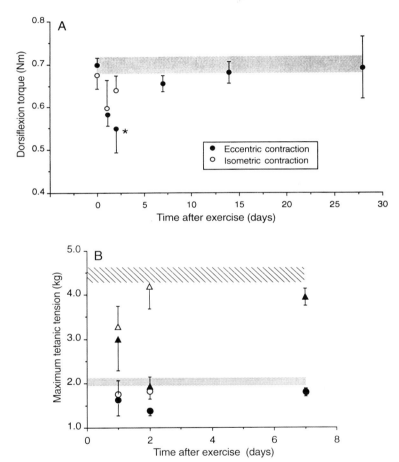

Fig. 3.11 (A) Time course of torque change after non-invasive eccentric contraction (filled circles) or non-invasive isometric contraction (open circles). Data represents mean ± SEM (n = 6 animals per time point). (From Leiber *et al.* 1994.) (B) Time course of changes in muscle force during the 7 days after eccentric contraction for the EDL (circles) and TA (triangles). Filled circles represent muscle forces measured after eccentric contraction. Open symbols represent muscle forces after contraction. Bars denote the normal values for TA and EDL. Note that the decline in muscle force is significantly greater for EDL than for TA.

decline. However, since torque continued to fall 2 days after eccentric contraction, at which stage muscles subjected to isometric contraction had recovered almost completely, eccentric contraction appeared to have triggered events that further decreased muscle performance. An obvious candidate is the inflammatory process that has already been implicated in muscle injury (Cannon *et al.* 1990; Evans and Cannon 1991).

3.4   Anti-inflammatory pharmacotherapy after muscle injury

Muscle fibre disruption, of the kind occurring after eccentric exercise, would be expected to constitute a significant inflammatory stimulus. An inflammatory process consequent on the initial injury was suggested as a possible explanation of the continued decrease in tension that followed the initial decline after eccentric contraction (Fig. 3.11(A)).

Evidence in support of an inflammatory process came from morphological examination of muscle fibres injured by repeated eccentric contraction. This showed that large fibres became thinner at variable distances from a given plane of cross-section. In the thick part of the fibre, myofibrils as well as the sarcolemma were in clear evidence; in the thinner portions myofibrils were widely separated and often absent. In sections distal or proximal to the thin portion, the fibre regained its contractile elements, and myofibrils became more densely packed. At the transitional region between thick and thin portions of a fibre, numerous inflammatory cells were found, although to a consistently smaller extent than in the thin region (Fridén *et al.* 1996).

Since the inflammatory process—which includes proteolysis by infiltrating neutrophils and macrophages—can cause damage in excess of that originally experienced by the tissue, it could be argued that prevention of inflammation would improve muscle status following injury. Such a hypothesis is difficult to test in humans, since the analgesic effect of non-steroidal anti-inflammatory drugs (NSAIDs) may itself permit of improved performance. We therefore examined the effect of NSAID treatment (with flurbiprofen) following eccentric contraction-induced muscle injury in our rabbit skeletal muscle model . Compared to a non-treated group that had been subjected to the same contraction regime, the NSAID-treated group demonstrated remarkable recovery of torque after 3 and 7 days, but then showed a significant decline in torque generation after 28 days (Fig. 3.12). This was an obvious feature in the torque records and also in force records obtained from both the EDL and TA. In support of this conclusion, two-way analysis of variance revealed a highly significant interaction between treatment and time ($p < 0.01$): muscles from flurbiprofen-treated animals were stronger than controls in the early period after eccentric contraction but weaker after 28 days (Mishra *et al.* 1995). Thus NSAID treatment appears to be beneficial in the short term but detrimental in the long term.

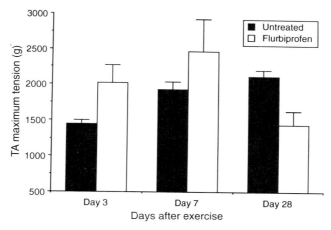

Fig. 3.12 Maximum tetanic tension of TA muscles from flurbiprofen-treated (filled bars) versus untreated animals. Tetanic tension mirrored the results obtained by measurement of dorsiflexion torque in that flurbiprofen-treated animals generated higher muscle forces early in the treatment and lower muscle forces later in the treatment.

3.5 Differential injury to TA and EDL muscles

To investigate in more detail the changes in torque observed after eccentric contraction, muscle tetanic tension was measured in both TA and EDL muscles—the major dorsiflexors of the ankle. Interestingly, we found that the magnitude of muscle damage was much greater in the EDL than the TA (Fig. 3.11(B)). For example, whereas the maximum tetanic tension developed by the TA decreased by about 40 per cent, that of EDL fell by almost 70 per cent! This was unlikely to be due to any specific differences in fibre-type distribution since the TA and EDL differ by only about 5 per cent in fibre type proportions. It is interesting to note that Salmons and his colleagues also observed more injury to the EDL than to the TA in a model based on stimulation-induced damage (see §5.9).

We therefore calculated the magnitude of the linear strain in each muscle that would accompany joint rotation under the conditions of the experiment. This was done by individual measurement of the moment arm for the TA and the EDL muscle according to the methods developed by An *et al.* (1983). Briefly, the distal tibia was secured via Steinmann pins to vertical braces. Another pin was placed through the talus and secured to a clamp which permitted ankle rotation. Stainless steel sutures were tied to the distal tendon stumps and then secured to toothed nylon cables connected to non-backlash gears mounted on potentiometers. In this way, translational movements of EDL and TA tendons could be related to a given joint rotation, and the moment arms for these muscles about the ankle calculated. Then, from a knowledge of the fibre length of each muscle

Muscle Damage

(Lieber and Blevins 1989), we calculated muscle fibre strain. The results are summarized in Fig. 3.13, where it can be seen that the magnitude of the tension decline appears to be a function of muscle fibre strain, since the muscles have nearly the same moment arm and thus experience approximately the same deformation during joint rotation. These data are therefore consistent with the idea that muscle strain causes injury.

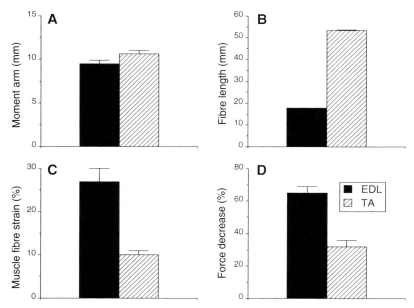

Fig. 3.13 Anatomical and physiological data from EDL (filled bars) and TA (hatched bars). Data represent mean ± SEM for $n = 6$ animals per group 2 days after eccentric contraction. (A) Moment arm; (B) fibre length; (C) muscle fibre strain; (D) muscle force decrease after eccentric contraction.

The implications of these findings for exercise involving eccentric contraction are not clear. Clearly it is not appropriate to conclude simply that low-amplitude joint excursions will result in low-strain muscle movements because, for a given joint rotation, length change varies from one muscle to another. Further studies of muscle-joint interaction will be required before isolated muscle strain measurements can be extrapolated to joint angle rotations in the intact individual.

3.6   Mechanics of muscle injury

In studying the mechanics of the injury process itself, we measured mechanical changes occurring within the first few minutes of eccentric contractions (Lieber *et al.* 1991). The fact that eccentric contractions are associated with high forces (Fig. 3.3(B)) and also result in muscle damage suggests the possibility that these high forces are the actual cause of injury. On the other hand, the differential injury to TA and EDL muscles considered in the preceding section led to the conclusion that strain was the important factor. To distinguish between these possibilities we designed an experiment in which maximum tetanic tension was measured after cyclic eccentric contractions in which muscle stress and strain were varied systematically; this was achieved by altering the timing of stimulation in relation to the muscle deformation in order to achieve different stresses at identical strains (Fig. 3.14). The experiment was repeated for different magnitudes of strain. We found that the extent of the force deficit was related to the magnitude of the muscle strain, rather than the stress imposed on the fibres (Fig. 3.15).

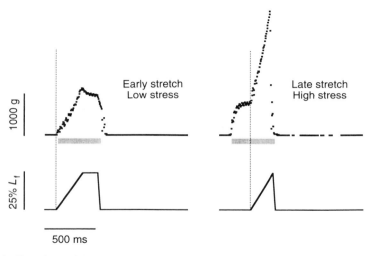

Fig. 3.14 Experimental design for varying stress and strain to determine mechanism of muscle injury. Lower panel, muscle deformation in terms of muscle fibre length (L_f). Upper panel, muscle tension. Horizontal stippled bar represents the period during which the skeletal muscle is stimulated. Vertical dotted lines represent the onset of muscle length change. Note that, by delaying the onset of muscle length change relative to stimulation, a high-stress condition is produced at the same strain. (Data from Lieber and Fridén 1993.)

3.7   Recent developments in mechanics and mechanisms of muscle injury

3.7.1   Mechanics of injury

Several recent studies add to our knowledge of the mechanism of muscle injury after eccentric contraction. For example, notwithstanding our finding that muscle strain, not stress, was the important determinant of muscle injury, an independent study performed on the rat soleus muscle (Warren *et al.* 1993) led to the conclusion that muscle stress was the more important factor. The reason for the discrepancy is not clear. However it is possibly relevant to note that the experiments in which strain was found to be the major determinant of muscle injury were performed on skeletal muscles composed mainly of fast-contracting fibres, whereas the study demonstrating that stress was the cause of muscle injury used a muscle composed primarily of slowly contracting fibres.

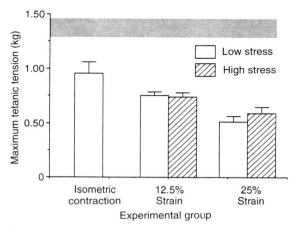

Fig. 3.15 Maximum tetanic tension (P_o) generated by rabbit TA muscles following cyclic eccentric contraction at various stresses and strains. Hatched bars represent P_o following high-stress cyclic eccentric contraction (created using stimulation pattern shown in Fig. 3.14). Open bars represent P_o following low-stress eccentric contraction. Note that the magnitude of the decline in tension after eccentric contraction depends on muscle strain, not muscle stress. Horizontal stippled area represents mean $\pm$ SEM ($n = 11$ for each group) for normal rabbit TA muscles. (Data from Lieber and Fridén 1993.)

A further critical difference between the studies was that qualitatively different experimental paradigms were used to impose the various force levels. We varied muscle force by altering the relative *timing* of stimulation and muscle stretch (Fig. 3.14). In contrast, Armstrong and his colleagues (Warren *et al.* 1993) varied stimulation *frequency* to change muscle force. Although both methods alter muscle force, muscle stiffness is quite different in the two cases. Alteration of the stimulation frequency results in high and low forces that are accompanied by high and low muscle stiffness. Thus, an identical deformation applied to muscles contracting at the different frequencies will result in different sarcomere strains since the deformation will be differentially absorbed by the muscle and the non-linearly compliant tendon. The definitive experiment to determine the relative importance of stress and strain requires that sarcomere length be measured under the different conditions so that actual sarcomere strain (and not strain calculated from tissue deformation) can be used as the independent variable.

3.7.2 Mechanisms of injury

Mechanical measurements made during eccentric contractions provided evidence that the injurious events occurred at a quite early stage of the protocol. We therefore tested the idea that desmin loss might actually occur at this early stage by examining muscles only 5 or 15 minutes after eccentric contraction (Lieber *et al.* 1996). To our surprise, desmin-negative cells were observed in the EDL after only 5 minutes! This affected only about 3 per cent of the EDL fibres, but this finding is clearly the earliest structural abnormality ever demonstrated in eccentrically exercised skeletal muscle.

The mechanism by which loss of desmin staining occurs is not known. However, we would speculate that the rapid change in desmin staining is not likely to be mediated by mechanisms that rely on gene regulation. Rather the finding points to mechanisms involving muscle fibre membrane disruption, with subsequent proteolysis or conformational change of the cytoskeletal network. A possible candidate for the proteolytic mechanism is the calcium-activated protease calpain, which is present in skeletal muscle (Dayton *et al.* 1976) and for which desmin is a substrate (Belcastro 1993; see also §6.2.2.1). The mechanism of action of calpain requires raised intracellular calcium ion ($[Ca^{2+}]_i$) concentration. Although there is, as yet, no direct evidence for such an increase, Duan *et al.* (1990) demonstrated an increase in the mitochondrial calcium concentration in muscles subjected to an exercise protocol biased towards eccentric contraction. Since mitochondrial calcium concentration indirectly reflects cytoplasmic $[Ca^{2+}]_i$, the results of Duan *et al.* might be construed as indirect evidence for increased $[Ca^{2+}]_i$. This observation, in conjunction with our earlier demonstration that muscle fibre strain was the mechanical factor that most strongly influenced the magnitude of muscle injury (Lieber and Fridén 1993), leads to the following hypothesis for the mechanism underlying the early muscle damage induced by eccentric contraction (Fig. 3.16).

(1) Muscle fibre strain results in an increase in $[Ca^{2+}]_i$. Such an increase may be due to calcium influx via strain-activated channels (Guharay and Sachs 1984), or disruption of the sarcoplasmic reticulum, T-system, or sarcolemma (Fig. 3.16(A)). This may be related to the concept of sarcomere 'popping' that has been proposed as a damage mechanism during eccentric contraction (Morgan 1990; also see §2.4.1).

(2) Following the increase in $[Ca^{2+}]_i$, calpain activation results in selective hydrolysis or disruption of the intermediate filament network (Fig. 3.16(B)). It has been demonstrated that desmin is a substrate for

Muscle Damage

calpain, whereas actin and myosin are not (Reddy *et al.* 1975). This could explain the loss of desmin in sections that still demonstrate a regular arrangement of contractile proteins.

(3) Finally, after the intermediate filament network has been altered by proteolysis or conformational change, the myofibrillar apparatus is disrupted on repeated muscle activation so that it is unable to develop normal tension (Fig. 3.16(C)).

Of course, numerous variations on this scheme could be proposed.

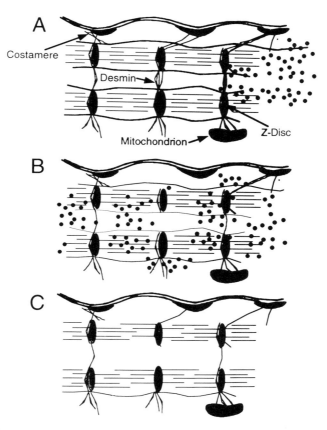

Fig. 3.16 Schematic depiction of eccentric contraction-induced muscle damage. Calcium ions are represented as filled dots. (A) Muscle fibre strain results in an increased $[Ca^{2+}]_i$. (B) Increased $[Ca^{2+}]_i$ leads to calpain activation and selective hydrolysis or disruption of the intermediate filament network. (C) Damage to the intermediate filament network results in disruption of the myofibrillar apparatus on repeated muscle activation.

3.8 Summary

When they are undergoing active lengthening, skeletal muscles possess unique mechanical properties. Although the biophysical basis of eccentric contractions is not completely understood, it is agreed that eccentric contractions are associated with muscle damage, soreness, and elevated serum enzyme levels. In the acute phase, muscles are more sensitive to strain than to stress. After injury has occurred, another process is triggered which results in further muscular deterioration. Future studies will be required to determine the precise events involved in this secondary process as well as to investigate methods for prevention of injury and promotion of recovery. Changes in the cytoskeleton are clearly involved in the manifestation of injury, which may be the direct result of cytoskeletal disruption or may represent an adaptive remodelling of the muscle fibre that occurs in response to mechanical overload. At all events, observation of changes in muscle proteins and morphology will undoubtedly lead to sensitive and specific methods for probing the cellular response to injury.

Acknowledgements

This work was supported by the Veterans Administration, NIH Grant AR40050, the Research Council of the Swedish Sports Federation, the Swedish Medical Research Council, and the Medical Faculty of the University of Umeå.

The authors acknowledge Professor Lars-Eric Thornell for providing antibodies and for interesting discussions. John Butler, Lena Carlsson, Chris Giangreco, Ulla Hedlund, Dev Mishra, Anna-Karin Nordlund, Mary Schmitz, Christy Trestik, and Abbe Zaro are thanked for technical assistance.

References

An, K.N., Ueba, Y., Chao, E.Y., Cooney, W.P., and Linscheid, R.L. (1983). Tendon excursion and moment arm of index finger muscles. *Journal of Biomechanics*, **16**, 419–25.

Armstrong, R.B., Ogilvie, R.W., and Schwane, J.A. (1983). Eccentric exercise-induced injury to rat skeletal muscle. *Journal of Applied Physiology*, **54**, 80–93.

Belcastro, A. (1993). Skeletal muscle calcium-activated neutral protease (Calpain) with exercise. *Journal of Applied Physiology*, **74**, 1381–6.

Cannon, J.G., Orencole, S.F., Fielding, R.A., Meydani, M., Meydani, S.N., Fiatarone, M.A., Blumberg, J.B., and Evans, W.J. (1990). Acute phase response in exercise: interaction of age and Vitamin E on neutrophils and muscle enzyme release. *American Journal of Physiology*, **259**, R1214–19.

Dayton, W.R., Goll, D.E., Zeece, M.G., Robson, R.M., and Reville, W.J. (1976). A Ca^{2+}-activated protease possibly involved in myofibrillar protein turnover purification from porcine muscle. *Biochemistry*, **15**, 2150–8.

Duan, C., Delp, M.D., Hayes, D.A., Delp, P.D., and Armstrong, R.B. (1990). Rat skeletal muscle mitochondrial (Ca^{2+}) and injury from downhill walking. *Journal of Applied Physiology*, **68**, 1241–51.

Evans, W.J. and Cannon, J.G. (1991). The metabolic effects of exercise-induced muscle damage. In *Exercise and sport sciences review*, pp. 99–125. Williams and Wilkins, Baltimore, Maryland.

Fridén, J., Sjöström, M., and Ekblom, B. (1981). A morphological study delayed muscle soreness. *Experientia*, **37**, 506–7.

Fridén, J., Sjöström, M., and Ekblom, B. (1983). Myofibrillar damage following intense eccentric exercise in man. *International Journal of Sports Medicine*, **4**, 170–6.

Fridén, J., Seger, J., and Ekblom, B. (1988). Sublethal muscle fibre injuries after high-tension anaerobic exercise. *European Journal of Applied Physiology and Occupational Physiology*, **57**, 360–8.

Funatsu, T., Higuchi, H., and Ishiwata, S. (1990). Elastic filaments in skeletal muscle revealed by selective removal of thin filaments with plasma gelsolin. *Journal of Cell Biology*, **110**, 53–62.

Gordon, A.M., Huxley, A.F., and Julian, F.J. (1966). The variation in isometric tension with sarcomere length in vertebrate muscle fibres. *Journal of Physiology (London)*, **184**, 170–92.

Goslow, G.J., Reinking, R., and Stuart, D. (1973). The cat step cycle: hind limb joint angles and muscle lengths during unrestrained locomotion. *Journal of Morphology*, **141**, 1–42.

Guharay, F. and Sachs, F. (1984). Stretch-activated single ion channel currents in tissue-cultured embryonic chick skeletal muscle. *Journal of Physiology (London)*, **352**, 685–701.

Harry, J.D., Ward, A.W., Heglund, N.C., Morgan, D.L., and McMahon, T.A. (1989). Crossbridge cycling theories cannot explain high speed lengthening behavior in frog muscle. *Biophysical Journal*, **57**, 201–8.

Horowits, R. and Podolsky, R.J. (1987). The positional stability of thick filaments in activated skeletal muscle depends on sarcomere length: evidence for the role of titin filaments. *Journal of Cell Biology*, **105**, 2217–23.

Horowits, R., Kempner, E.S., Bisher, M.E., and Podolsky, R.J. (1986). A physiological role for titin and nebulin in skeletal muscle. *Nature (London)*, **323**, 160–4.

Katz, B. (1939). The relation between force and speed in muscular contraction. *Journal of Physiology (London)*, **96**, 45–64.

Kuipers, H., Drukker, J., Frederik, P.M., Geurten, P., and Kranenburg, G.V. (1983). Muscle degeneration after exercise in rats. *International Journal of Sports Medicine*, **4**, 45–51.

Lazarides, E. (1982). Intermediate filaments: a chemically heterogeneous, developmentally regulated class of proteins. *Annual Review of Biochemistry*, **51**, 219–50.

Lieber, R.L. and Blevins, F.T. (1989). Skeletal muscle architecture of the rabbit hindlimb: functional implications of muscle design. *Journal of Morphology*, **199**, 93–101.

Lieber, R.L. and Fridén, J. (1993). Muscle damage is not a function of muscle force but active muscle strain. *Journal of Applied Physiology*, **74**, 520–6.

Lieber, R.L., McKee-Woodburn, T., and Fridén, J. (1991). Muscle damage induced by eccentric contractions of 25% strain. *Journal of Applied Physiology*, **70**, 2498–507.

Lieber, R.L. (1992). *Skeletal muscle structure and function*. Williams and Williams, Baltimore, MD.

Lieber, R.L., Schmitz, M.C., Mishra, D.K., and Fridén, J. (1994). Contractile and cellular remodeling in rabbit skeletal muscle after cyclic eccentric contractions. *Journal of Applied Physiology*, **77**, 1926–34.

Lieber, R.L., Thornell, L.-E., and Fridén, J. (1996). Muscle cytoskeletal disruption occurs within the first 15 minutes of cyclic eccentric contraction. *Journal of Applied Physiology*, **80**, 278–84.

Magid, A. and Law, D.J. (1985). Myofibrils bear most of the resting tension in frog skeletal muscle. *Science*, **230**, 1280–2.

Mastaglia, F.L. and Walton, J. (1982). *Skeletal muscle pathology*. Churchill Livingstone, New York.

McCully, K.K. and Faulkner, J.A. (1985). Injury to skeletal muscle fibres of mice following lengthening contractions. *Journal of Applied Physiology*, **59**, 119–26.

Mishra, D.K., Fridén, J., Schmitz, M.C., and Lieber, R.L. (1995). Antiinflammatory medication after muscle injury. A treatment resulting in short-term improvement but subsequent loss of muscle function. *Journal of Bone and Joint Surgery. (American Volume)*, **77**, 1510–19.

Morgan, D.L. (1990). New insights into the behavior of muscle during active lengthening. *Biophysical Journal*, **57**, 209–21.

Newham, D.H., McPhail, G., Mills, K.R., and Edwards, R.H.T. (1983). Ultrastructural changes after concentric and eccentric contractions of human muscle. *Journal of Neurological Sciences*, **61**, 109–22.

Price, M. and Sanger, J.W. (1979). Intermediate filaments connect Z-discs in adult chicken muscle. *Journal of Experimental Zoology*, **208**, 263–9.

Reddy, M.K., Etlinger, J.D., Rabinowitz, M., Fischman, D.A., and Zak, R. (1975). Removal of Z-lines and a-actinin from isolated myofibrils by a calcium-activated neutral protease. *Journal of Biological Chemistry*, **250**, 4278–84.

Sanes, J.R. and Cheney, J.M. (1982). Laminin, fibronectin and collagen in synaptic and extrasynaptic portions of muscle fibre basement membrane. *Journal of Cell Biology*, **93**, 442–51.

Sternberger, L.A. (1979). *Immunocytochemistry*. Wiley Medical, New York.

Thornell, L.-E., Eriksson, A., Johansson, B., Kjorell, U., Franke, W.W., Virtanen, I. *et al.* (1985). Intermediate filament and associated proteins in heart purkinje fibres: a membrane-myofibril anchored cytoskeletal system. *Annals of the New York Academy of Sciences*, **455**, 213–40.

Tokuyasu, K.T., Dutton, A.H., and Singer, S.J. (1982). Immunoelectron microscopic studies of desmin (skeletin) localization and intermediate filament organization in chicken skeletal muscle. *Journal of Cell Biology*, **96**, 1727–35.

Vartio, T., Laitinen, L., Närvänen, O., Cutolo, M., Thornell, L.-E., and Virtanen, I. (1987). Differential expression of the ED sequence-containing form of cellular fibronectin in embryonic and adult human tissues. *Journal of Cell Science*, **88**, 419–30.

Wang, K. and Ramirez-Mitchell, R. (1983). A network of transverse and longitudinal intermediate filaments is associated with sarcomeres of adult vertebrate skeletal muscle. *Journal of Cell Biology*, **96**, 562–70.

Warren, G.W., Hayes, D., Lowe, D.A., and Armstrong, R.B. (1993). Mechanical factors in the initiation of eccentric contraction-induced injury in rat soleus muscle. *Journal of Physiology (London)*, **464**, 457–75.

4

Human muscle damage induced by eccentric exercise or reperfusion injury: a common mechanism?

David A. Jones and Joan M. Round

4.1 Introduction

In this chapter we will discuss some of the unexplained features of damage to human skeletal muscle caused by eccentric exercise and make comparisons with the rhabdomyolysis that can occur following the surgical repair and reperfusion of ischaemic limbs.

Despite the considerable attention devoted to the subject in the last 10 years, there are two major consequences of eccentric exercise that remain unexplained. The first is the origin of the characteristic muscle tenderness and pain, sensations that affect us all after unaccustomed or particularly heavy exercise. The second is the nature of the mechanism that gives rise to the delayed damage.

The salient features of the damage that occurs when human muscle is subjected to eccentric exercise are, in the order in which they occur:

(1) a loss of strength and a change in the force-frequency relationship of the muscle;
(2) pain and muscle tenderness, which reaches a peak 24–48 h after the exercise;
(3) release of soluble proteins (for example, creatine kinase (CK)) into the circulation, reaching a peak at about 5–6 days;
(4) destruction of muscle fibres, accompanied by infiltration of the tissue by mononuclear cells, between 6 and 10 days after exercise;
(5) activation of satellite cells, with the formation of myotubes and restoration of normal structure and function within 2 to 3 weeks.

We will tend to focus here on details, and particularly the effects of training, so for a wider view of the effects of eccentric exercise on skeletal muscle the reader is referred to Chapters 1, 2, and 3 of this book and studies by Asmussen (1953, 1956), Fridén *et al.* (1983), Newham *et al.* (1983*a, b*), and Jones *et al.* (1986). It should be added that the terms 'eccentric exercise' or 'eccentric contractions' will be used, without apology, to describe situations in which active muscles are extended, such as when lowering the body weight during stepping or walking downhill, or in experimental situations when muscles such as the biceps are forcibly extended by a modern version of the medieval rack (approval having first been obtained from the relevant Ethical Committee).

4.2   Muscle pain

During eccentric exercise the subject is aware of high forces generated in the muscle and tendons but not of the burning sensation due to metabolic changes that is associated with high-intensity isometric or concentric activity. After 10–20 min of eccentric exercise the subject usually notices problems in controlling the movement, a loss of force and increased tremor, and difficulty in flexing and extending the affected limb fully. These sensations are generally described as 'unusual', but rarely as 'painful'. Over the next 6–12 h, however, discomfort develops in the exercised muscles; typically the subject first notices pain, which can be acute, the morning after the exercise. The major sensation is one of muscle tenderness, a feeling of discomfort elicited by pressure. The feeling is similar, if not identical, to that of a bruise or sprain. At rest, and with nothing touching the muscle, there is no discomfort, but external pressure and stretching can be acutely painful. The sensation may be quantified by measuring the pressure required to elicit a painful response (Edwards *et al.* 1981) and this measurement reveals the characteristic time course shown in Fig. 4.1. Associated with the tenderness is a sensation of stiffness, which limits the range of movement of the limb (Jones *et al.* 1987*a*), together with signs of oedema over the affected muscle. Although the early suggestion was that the pain is associated with damage to connective tissue (Abraham 1977), the discovery of the considerable muscle fibre damage that can occur as a result of eccentric exercise focused attention on the muscle fibre as the source of algesic substances or the possibility of local muscle spasm triggered by damage to muscle fibre membranes (De Vries 1966). Local muscle contraction has now been shown not to be an important factor, as no unusual electrical activity can be recorded from the painful muscle (Howell *et al.* 1986; Jones *et al.* 1987*b*).

   Although the idea of algesic substances being released from damaged fibres is superficially attractive, there are three lines of evidence against this mechanism (Jones *et al.* 1987*b*). The first is that the time course of the development of pain and the release of soluble substances from the muscle fibres is different, the release of CK having barely begun at a time when the peak pain is experienced (compare Fig. 4.1(A) and (B)). The second line of evidence arises from the fact that the pain and CK responses are very variable, both between subjects and on different occasions, and, although

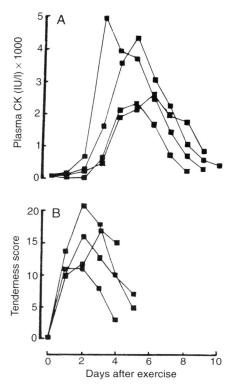

Fig. 4.1 Plasma CK and muscle tenderness after eccentric exercise. Four subjects performed 20 min of maximum eccentric exercise (once every 15 s) of the elbow flexors, and plasma CK and muscle tenderness were measured at daily intervals thereafter.

it is true that whenever a large CK response is seen it is always preceded by very painful muscles, the reverse does not necessarily hold. This difference could arise if people varied greatly in their perception of pain, but there is a third strand of evidence that argues against this objection. Figure 4.2 shows the effect of training once a week on the pain and CK responses of one subject. It can be seen that, whereas the CK response was much diminished after the first bout of exercise and absent thereafter, the pain was trained away more slowly so that by the third week, when there was still appreciable pain, there was no evidence of release of soluble protein from the muscle fibres. It is still possible that there is another factor, which is released from damaged muscle with a different time course and does not respond to training in the same way as CK, but this seems to be clutching at straws.

We see, therefore, that pain and muscle fibre damage are dissociated in time, the magnitudes of the two responses are not necessarily correlated, and, although both are reduced by training, they decline with different time courses. Taken together, these observations suggest that muscle tenderness after eccentric exercise is not caused by damage to muscle fibres and attention must focus on other elements within the muscle, such as the connective tissue components, as suggested originally by Abraham (1977).

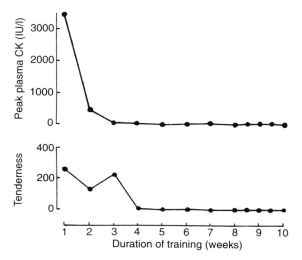

Fig. 4.2 Plasma CK and quadriceps tenderness with repeated exercise. One subject performed 20 min of stepping exercise once a week for 10 weeks. Peak values for plasma CK and muscle pain are shown for each week.

4.3   Muscle fibre damage

The most intriguing feature of the muscle damage arising from eccentric exercise is the delay between the end of the precipitating exercise and the evidence of damage in terms of the leakage of soluble constituents from the muscle fibres (Fig. 4.1). A delay of some hours might be explicable in terms of a slow diffusion of large proteins into the general circulation, as is seen with the appearance of markers of cardiac damage after an infarction. After eccentric exercise, however, there is usually a delay of about 2 days before any change is seen, and the peak response is seen 4 to 6 days after the exercise. Observations made on patients undergoing orthopaedic surgery in which large amounts of muscle are cut, or at least roughly handled, have shown relatively small increases in plasma CK that reached peak levels 1 or 2 days after the operation (Jones *et al.* 1991), so there is a clear difference between the time course of damage caused by eccentric exercise and that resulting from direct physical trauma to muscle fibres (Fig. 4.3). The delay between exercise and damage could be explained if the exercise caused an initial change in the muscle that was amplified over the next few days until it resulted in muscle necrosis. There are a number of changes in the structure and contractile characteristics of the muscle that may be indicators of this initial damage. Immediately after the exercise, changes can be seen in the ultrastructure of the muscle, such as Z-disc streaming and sarcomeres that have the appearance of being pulled apart. These changes may be limited to isolated sarcomeres, or extend to wider areas of disruption (Fridén *et al.* 1983; Newham *et al.* 1983c). Immediately after eccentric exercise there is a loss of maximum voluntary isometric force that has nothing to do with pain or loss of central drive because similar force loss can be shown by stimulating the muscle at a high frequency. An even greater loss of force is seen at low frequencies of stimulation, so the force-frequency curve is shifted to the right (Fig. 4.4; Newham *et al.* 1983b). The changes in strength and contractile characteristics persist for hours and sometimes days. Long-lasting alterations in function are generally ascribed to 'damage' rather than 'fatigue' since the latter carries with it the implication that recovery will occur within a period of minutes rather than hours. These long-lasting changes have been taken as evidence of 'micro' damage to the muscle fibres which, if of sufficient magnitude, could have a cumulative effect, overwhelming the capacity for repair and leading to necrosis of a segment, or the whole of the muscle fibre. However, just as it can be shown that pain and muscle fibre damage have different underlying mechanisms, so we will show that the 'micro' damage is not an adequate explanation for the subsequent muscle degeneration.

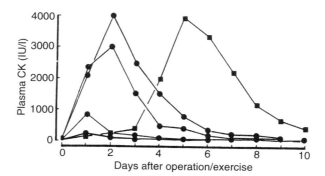

Fig. 4.3 Plasma CK measured in patients undergoing a range of orthopaedic operations. The values are compared to those obtained from one subject (filled squares) who had undertaken eccentric exercise of the biceps (as in Fig. 4.1).

One of the first descriptions of Z-disc streaming and sarcomere disruption as a consequence of eccentric exercise came from an examination of biopsies from the quadriceps muscles of subjects after stepping exercise (Newham *et al.* 1983c). It was noted, but not commented upon at the time, that there

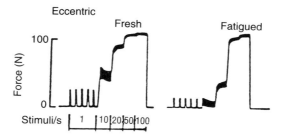

Fig. 4.4 Force-frequency relationships of fresh and exercised muscle. Force generated by percutaneous electrical stimulation of a portion of the quadriceps at 1, 10, 20, 50, and 100 Hz before, and 10 min after, a 20-min period of stepping. See Fig. 4.6 for an example of the changes in maximum force-generating capacity of the muscle that result from this type of exercise. Results are shown for the leg that had been used for stepping down.

was no correlation between the extent of damage seen with electron microscopy and the subsequent CK release. In a further series of experiments, radioactive technetium pyrophosphate was used to identify damaged muscle. These experiments showed that, after stepping exercise, the gluteus and one of the adductor muscles, but not the quadriceps, took up the isotope and was, therefore, the most probable site of CK release (Fig. 4.5; Newham *et al.* 1986). Thus there is evidence of microscopic damage to the quadriceps that does not necessarily progress to full-scale degeneration of the muscle fibres (although it is still possible that gross muscle damage ensues only if some threshold is exceeded).

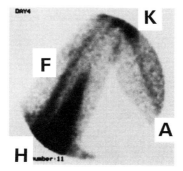

Fig. 4.5 Uptake of radioactive technetium pyrophosphate into muscle damaged by stepping exercise. Lateral view of the leg used to step down during the exercise, showing uptake into the adductor and gluteal muscles but no sign of uptake into the quadriceps. A, ankle; H, hip; F, femur; K, knee.

The Z-disc streaming and disruption of sarcomere structure probably includes damage to the sarcoplasmic reticulum and T-tubule systems and it is reasonable to assume that the immediate loss of force and change in the force-frequency relationship are reflections of the extent of this damage. It has been noted that the changes in contractile function do not correlate well with the extent of subsequent fibre degeneration, indicated by CK release. This inconsistency is illustrated well by considering the effects of training. Figure 4.6 shows the loss of strength after a series of bouts of exercise at two-weekly intervals (Newham *et al.* 1987). On each occasion the initial loss of force approached 50 per cent, with recovery taking a little over a week, and a similar pattern of change was seen in the force-frequency relationships. Figure 4.7 shows the very different pattern of CK release from the exercised muscle between the first and the two subsequent bouts of exercise. Following the first exercise session there was a large release of CK into the circulation, with individual values ranging from 1570 to 10 904 IU/l (upper limit of normal, 200 IU/l), which reached a peak at 5 days and had returned to basal levels within 10 days. Following the second and third exercise sessions there was no significant increase in circulating CK, indicating that the muscle had become resistant to the destructive process. The contrast to the changes in the contractile properties of the muscle under the same conditions (Fig. 4.6) shows quite clearly that, if 'micro' damage is responsible for the changes in contractile properties, it is unlikely to be a precursor of the delayed onset muscle degeneration, especially as the extent of the initial loss of force was so similar on the three occasions.

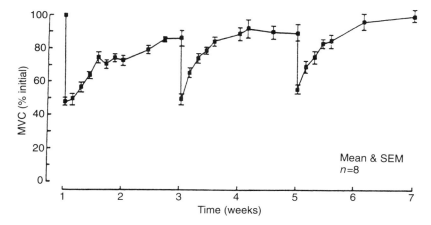

Fig. 4.6 Changes in strength with repeated exercise. Eight subjects exercised their forearm flexors with maximal eccentric contractions as described in the legend to Fig. 4.1, repeating the exercise every 2 weeks. Maximum isometric strength was measured at regular intervals and the results are expressed as a percentage of the strength before the first period of exercise. MVC, maximal voluntary contraction.

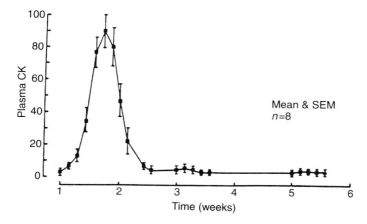

Fig. 4.7 Changes in CK response to repeated exercise. Plasma CK measurements were made on the subjects described in Fig. 4.6 and the results are expressed as a percentage of the highest value obtained for each individual.

So far we have excluded a number of possible causes of muscle fibre damage but we have not suggested alternative mechanisms. It may be instructive, therefore, to consider a different type of muscle damage to see whether it provides any parallels that could throw light on exercise-induced pain and damage.

4.4 Reperfusion injury to skeletal muscle

The techniques of vascular surgery are now so well advanced that it is possible to salvage limbs which several years ago would have been amputated. One of the side-effects of these advances has been the occasional case in which, following successful revascularization, the patient experiences severe rhabdomyolysis and subsequent renal failure.

Figure 4.8 follows the appearance of CK and myoglobin in the plasma of one patient who had undergone a technically successful repair of the femoral artery (Adiseshiah *et al.* 1992). The plasma creatinine shows the severity of the renal complications, which necessitated prolonged dialysis and were caused by accumulation of myoglobin in the renal tubules. In this type of case, immediately after surgery the affected limb becomes pink and warm with every sign of a successful outcome but within hours a marked oedema develops which can become so severe as to endanger the circulation and to require a fasciotomy to relieve the pressure. Immediately after revascularization there is little increase in circulating CK and the first pass of blood through the muscle does not flush out any significant CK activity. Therefore the large CK efflux that occurs subsequently must be a result of degenerative changes that occurred after the muscle had been revascularized, rather than a release of soluble constituents during the ischaemic period. Affected muscles become tender and very sensitive to touch. Muscle samples obtained either during or immediately after surgery (Fig. 4.9(A)) show little or no evidence of damage at the light microscopic level (although it would be most interesting to know what might be revealed by electron microscopic examination at this stage). Several days later, however, there is marked muscle fibre necrosis and infiltration by mononuclear cells (Fig. 4.9(B)), an appearance very similar to that of a muscle damaged by eccentric exercise (Fig. 4.9(C)).

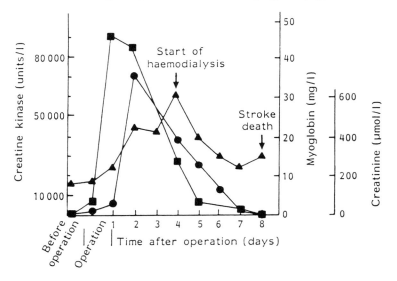

Fig. 4.8   Reperfusion injury to muscle. Results from a patient operated on to repair damage to the femoral artery of one leg. Values are given for plasma levels of myoglobin (filled squares), CK (filled circles), and creatinine (filled triangles).

The mechanism underlying the muscle damage following reperfusion is unclear but there are suggestions that free radical-mediated processes are involved, radicals being generated when oxygenated blood returns to the metabolically depleted tissue. In a number of tissues the conversion of xanthine dehydrogenase to xanthine oxidase has been implicated in the process of reperfusion injury. A similar mechanism might be suggested for skeletal muscle but for the fact that xanthine oxidase is not found in skeletal muscle fibres (McCord 1985; see also §6.2.3). However, the enzyme is found in the capillary endothelium and it is possible that the endothelium is the site of the primary damage in reperfusion injury. The rhabdomyolysis may be a secondary phenomenon consequent upon the increased permeability of the capillaries.

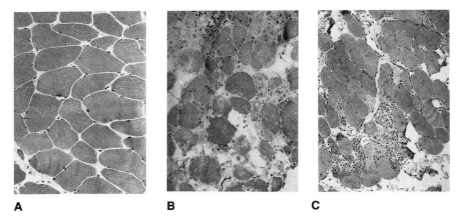

A   **B**   **C**

Fig. 4.9 Histological appearance of damaged muscle. (A) and (B) Sections from muscle biopsies taken from the tibialis anterior muscle of a patient whose case was similar to that described in Fig. 4.8. (A) Sample taken during the operation; (B) sample taken 10 days after the operation; (C) biopsy from the biceps of a normal subject taken 10 days after a period of eccentric exercise. Sections stained with haematoxylin and eosin.

4.5　Summary and conclusions

The large CK release that can occur several days after eccentric exercise is unlikely to be a consequence of major physical trauma to the muscle during the exercise, because direct injury to muscle—during orthopaedic surgery, for example—results in a comparatively rapid release of soluble constituents. According to the concept of 'micro' damage, the initial exercise causes some small amount of damage which multiplies until, several days later, it reaches critical proportions; the fibre, or a segment of it, then succumbs, undergoing autolysis and then regeneration. The signs of 'micro' damage are, first, direct ultrastructural evidence of sarcomere disruption and, secondly, changes in the strength and contractile properties of the muscle. However, as discussed above, these signs do not correlate well with the subsequent muscle fibre destruction. In particular, training has a dramatic effect on the release of CK from muscles yet it has little effect on the changes in contractile properties which are signs of the supposed precursor events leading to the delayed onset muscle damage.

It is possible that free-radical damage to the capillary endothelium is the factor that initiates the severe rhabdomyolysis that can occur after reperfusion of an ischaemic limb. It occurs to us that disruption of the capillary endothelium by mechanical trauma could be the precipitating factor in the muscle damage caused by eccentric exercise. The outward signs of the two forms of damage are very similar: oedema, tenderness, release of soluble

proteins, and extensive muscle fibre necrosis with infiltration by mononuclear cells. The major difference lies in the time course: reperfusion injury
is relatively rapid whereas injury following eccentric exercise is delayed. This
difference could reflect the different causes of damage to the endothelium:
free radicals in the one case and physical trauma in the other. We have no
suggestions as to how capillaries could be damaged by eccentric contractions.
Stretch alone is unlikely to be responsible, since passive stretching of a muscle
does not result in damage. However, during eccentric contractions high
forces, pressures, and shearing forces are developed within the muscle that
could disrupt the smaller vessels. Any proposed mechanism must explain the
effects of training: the capillary bed of a muscle is known to be very adaptable
to endurance training, although we have no specific proposals as to how it
might respond to repeated bouts of eccentric exercise.

At present there is a wealth of information about the circumstances and
time course of muscle damage and the associated phenomena caused by
eccentric exercise. There is, however, a dearth of ideas about the specific
mechanism that links this one form of exercise to its characteristic sequelae.
There is circumstantial evidence to suggest that damage to the capillary
endothelium could be an important factor in the development of muscle pain
and/or delayed onset muscle damage and this hypothesis should be amenable
to experimental investigation.

Acknowledgements

The authors have received generous financial support from The Wellcome
Trust, Action Research, and Research into Ageing.

References

Abraham, W.M. (1977). Factors in delayed muscle soreness. *Medicine and Science in
Sports*, **9**, 11–26.
Adiseshiah, M., Round, J.M., and Jones, D.A. (1992). Reperfusion injury in skeletal
muscle: a prospective study in patients with acute limb ischaemia and claudicants
treated with revascularization. *British Journal of Surgery*, **79**, 1026–9.
Asmussen, E. (1953). Positive and negative muscular work. *Acta Physiologica
Scandinavica*, **28**, 364–82.
Asmussen, E. (1956). Observations on experimental muscle soreness. *Acta Rheuma-
tologica Scandinavica*, **2**, 109–16.
De Vries, H.A. (1966). Quantitative electromyographic investigation of the spasm
theory of muscle pain. *Journal of Physical Medicine*, **45**, 119–34.
Edwards, R.H.T., Mills, K.R., and Newham, D.J. (1981). Measurement of severity
and distribution of experimental muscle tenderness. *Journal of Physiology*, **317**,
1–2P.

Fridén, J., Sjöström, M., and Ekblom, B. (1983). Myofibrillar damage following intense eccentric exercise in man. *International Journal of Sports Medicine*, **4**, 170–6.

Howell, J.H., Chila, A.G., Ford, G., David, D., and Gates, T. (1986). An electromyographic study of elbow motion during postexercise muscle soreness. *Journal of Applied Physiology*, **58**, 1713–18.

Jones, D.A., Newham, D.J., Round, J.M., and Tolfree, S.E.J. (1986). Experimental human muscle damage: morphological changes in relation to other indices of damage. *Journal of Physiology*, **375**, 435–48.

Jones, D.A., Newham, D.J., and Clarkson, P.M. (1987a). Skeletal muscle stiffness and pain following eccentric exercise of the elbow flexors. *Pain*, **30**, 233–42.

Jones, D.A., Newham, D.J., Obletter, G., and Giamberadino, M.A. (1987b). Nature of exercise-induced muscle pain. *Advances in Pain Research*, **10**, 207–18.

Jones, D.A., Round, J.M., and Carli, F. (1991). Plasma creatine kinase in patients following routine surgery: a comparison with experimental muscle damage. *Journal of Physiology*, **438**, 173P.

McCord, J.M. (1985). Oxygen-derived free radicals in post ischaemic tissue injury. *New England Journal of Medicine*, **312**, 159–63.

Newham, D.J., Jones, D.A., and Edwards, R.H.T. (1983a). Large and delayed creatine kinase changes after stretching exercise. *Muscle and Nerve*, 6, 36–41.

Newham, D.J., Mills, K.R., Quigley, B.M., and Edwards, R.H.T. (1983b). Pain and fatigue following concentric and eccentric muscle contractions. *Clinical Science*, **64**, 55–62.

Newham, D.J., McPhail, G., Mills, K.R., and Edwards, R.H.T. (1983c). Ultrastructural changes after concentric and eccentric contractions. *Journal of the Neurological Sciences*, **61**, 109–22.

Newham, D.J., Jones, D.A., Tolfree, S.E.J., and Edwards, R.H.T. (1986). Skeletal muscle damage: a study of isotope uptake, enzyme efflux and pain after stepping. *European Journal of Applied Physiology*, **55**, 106–12.

Newham, D.J., Jones, D.A., and Clarkson, P.M. (1987). Repeated high force eccentric exercise: effects on muscle pain and damage. *Journal of Applied Physiology*, **63**, 1381–6.

5

Muscle damage induced by neuromuscular stimulation

Jan Lexell, Jonathan C. Jarvis, David Y. Downham,
and Stanley Salmons

5.1   Background

Plasticity of mammalian skeletal muscle was first manifested in experiments in which the motor nerves of different muscles were cross-anastomosed (Buller *et al.* 1960). With the introduction of the implantable muscle stimulator (Salmons 1967) it became possible to show that this phenomenon was related to changes in use: merely modifying the impulse activity in the motor nerve over an extended period produced changes in the corresponding muscle fibres that were even more profound than those resulting from nerve cross-union (Salmons and Vrbová 1969; Salmons and Sréter 1976). In particular, fibres of the fast-twitch, fatigue-susceptible type that were subjected to sustained high levels of use for several weeks developed slow-contracting, fatigue-resistant properties.

Since adult mammalian muscle was regarded as a terminally differentiated tissue, the scope and extent of these changes was unexpected, and initial reports of the phenomenon (Salmons and Vrbová 1969) were not readily accepted. Attempts were made to explain the type transformation as the result of selective degeneration of fast muscle fibres and proliferation of slow muscle fibres. However, a series of studies provided convincing evidence that: (1) changes in response to stimulation took place within pre-existing fibres (Pette and Schnez 1977; Rubinstein *et al.* 1978; Salmons and Henriksson 1981); and (2) a turnover of fibre populations, which would have been accompanied by large-scale degeneration and regeneration, did not occur (Salmons and Henriksson 1981; W.E. Brown *et al.* 1983; Eisenberg *et al.* 1984; Hoffman *et al.* 1985; J.M.C. Brown *et al.* 1989). It was concluded, and generally accepted, that the destruction of pre-existing fibres and their replacement by regeneration were not obligatory parts of the transformation process. The phenomenon could be interpreted as an adaptive response (Salmons and Henriksson 1981): changes induced by chronic stimulation in the mechanochemistry of contraction, the kinetics of calcium transport and storage, and the metabolic pathways for the generation of ATP enabled the muscle to support sustained work without fatigue by decreasing the energy needed for contraction and increasing the capacity for generating that energy through aerobic routes. This remarkable capacity has awakened interest in the surgical potential for grafts of fatigue-resistant muscle in, for example, cardiac assistance and the treatment of incontinence (see §11.1).

More recently, the suggestion that degenerative and regenerative phenomena play a significant role in muscle type transformation was revived (Maier *et al.* 1986, 1988). These authors provided evidence of stimulation-induced degeneration involving as much as a quarter of the muscle fibre population. The observations were difficult to reconcile with evidence, already referred to, that had accumulated to the contrary, and were in direct conflict with earlier studies by Eisenberg and Salmons (1981), who did not detect, at either the light- or the electron-microscopic level, signs of degenerative or regenerative change resulting from stimulation for periods ranging from 6 hours to 5 months.

Is the destruction of pre-existing fibres and their replacement by regeneration an obligatory part of the transformation process? The great bulk of the evidence argues against such a proposition. But, even if it is not essential to transformation, is a degree of muscle damage an inevitable concomitant of chronic indirect stimulation? The question has serious clinical implications. Damage induced in the early stages of stimulation would in all probability be repaired by regeneration in the long term, but the resultant delay in realizing the full functional potential of the muscle would be highly undesirable from a therapeutic point of view. Ongoing episodes of damage, even at a lower level, would be equally disturbing. We therefore need to understand how and to what extent damage is induced by electrical stimulation if the therapeutic possibilities of functional neuromuscular stimulation are to be exploited.

We started by examining systematically those differences in experimental design and technique that could underlie disparities in the reported incidence of stimulation-induced muscle damage (Lexell *et al.* 1992, 1993). For this purpose, light microscopical techniques were combined with a statistical approach to the quantitative assessment of damage. The protocol takes account of different sources of variation, such as the effects of sampling, the choice of muscle, and the pattern and frequency of stimulation. The results reviewed here explain many of the conflicts in the literature.

5.2   Stimulation

A total of 26 New Zealand White rabbits with body weights in the range 2.5–3.5 kg were used. Two rabbits served as 'unoperated controls'. In the other 24 rabbits, an implantable stimulator of established design (Jarvis and Salmons 1991; Salmons and Jarvis 1991) was placed under the skin of the left flank and the electrodes were sutured close to the common peroneal nerve.

Each stimulator had an integral power source and a circuit that allowed remote switching via a transcutaneous optical link.

The 24 rabbits were divided into six groups. In the first group ($n = 5$), the common peroneal nerve was stimulated at 10 Hz for 24 h/day ('10 Hz continuous'). In the second group ($n = 5$), the 10 Hz pattern was delivered for one hour 'on' and one hour 'off' throughout the day, a total of 12 h/day ('10 Hz intermittent'). In the third ($n = 4$), fourth ($n = 4$), and fifth ($n = 4$) groups, the common peroneal nerve was stimulated at 5, 2.5, and 1.25 Hz, respectively, for 24 h/day ('5 Hz continuous'; '2.5 Hz continuous'; '1.25 Hz continuous'). All stimulating pulses had an amplitude of 3 V and a pulse duration of 200 μs. Stimulation was initiated at least 2 weeks after the operation and lasted 9 days. The two remaining rabbits served as 'operated, non-stimulated controls', so that there were two operated and two non-operated control rabbits.

All animal procedures were performed in strict accordance with the Animals (Scientific Procedures) Act 1986, which governs the humane care and use of laboratory animals in the United Kingdom.

5.3   Tissue processing and general morphological assessment

The tibialis anterior (TA) and extensor digitorum longus (EDL) muscles of the left and the right hindlimbs were removed from every rabbit. Each muscle was weighed and cut into 10–12 transverse slices. The slices were frozen either in liquid nitrogen or in isopentane chilled with liquid nitrogen. Serial cryosections of 5–10 μm thickness were cut and stained by the following procedures: haematoxylin and eosin (H & E) and modified Gomori trichrome, for general morphological assessment; Verhoeff-van Gieson's stain, for connective tissue; succinate dehydrogenase, for the integrity and distribution of mitochondria; acid phosphatase, as a marker of activated macrophages and lysosomes; acridine orange, for nucleic acids. Sections were also reacted with a monoclonal antibody (WB-MHCn) against rabbit neonatal myosin heavy chain isoform (anti-neonatal), as a marker of newly formed fibres. Binding of the primary antibody was detected by immunofluorescence of a second antibody, rabbit anti-mouse immunoglobulin, conjugated with fluorescein isothiocyanate (FITC), and the sections were viewed with a Leitz Diaplan microscope fitted for epifluorescent illumination.

Where stimulation-induced damage occurred, its nature was not in question: the various staining techniques clearly revealed focal and diffuse signs indicative of degeneration as well as regeneration. In the affected areas, the internal architecture of fibres was disturbed; activated macrophages and lysosomes were present within and between fibres; the amount of RNA within fibres was increased; and some fibres reacted positively with a monoclonal antibody against the neonatal isoform of myosin.

The incidence of degeneration varied considerably between experiments, between TA and EDL muscles, and between rabbits within the same experimental group. In general, degenerative features were both less marked and less frequent at the lower frequencies (Fig. 5.1(A)-(C)), and were less prominent in TA muscles than in EDL muscles. Muscles in which the other staining techniques had revealed most evidence of degeneration showed the largest numbers of fibres reacting positively with the anti-neonatal antibody (Fig. 5.1(D)-(F)).

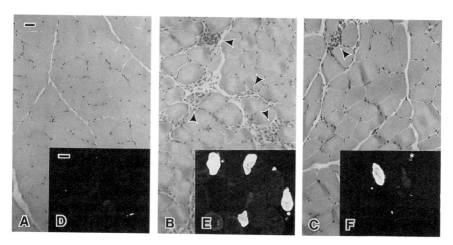

Fig. 5.1 Transverse cryostat sections of rabbit tibialis anterior muscle: (A)-(C) stained by haematoxylin and eosin; and (D)-(F) reacted with a monoclonal antibody against rabbit neonatal myosin heavy chain isoform. The series illustrates the incidence of degeneration and regeneration after 9 days of stimulation with different patterns and frequencies. (A) Control. (B) 10 Hz continuous stimulation. Fibres invaded by mononuclear cells and with disrupted cytoplasm, as well as clearly degenerated fibres, are seen (see arrows). (C) 2.5 Hz continuous stimulation. One area with foreign mononuclear cells between intact muscle fibres is seen (see arrow). (D) Control. (E) 10 Hz continuous stimulation. Four fibres that reacted positively with the anti-neonatal antibody are seen. (F) 2.5 Hz continuous stimulation. One fibre that reacted positively with the anti-neonatal antibody is seen. Scale bar, 25 μm; magnification, 100×.

5.4   Sampling procedure and statistical analyses

The results obtained with the different staining techniques were in excellent agreement. It was therefore considered sufficient to perform the measurements on sections stained with H & E. All sampling and quantification were done on 3–5 sections from each muscle, prepared from the midbelly of the muscle, and from slices 10 and 20 mm proximal and 10 and 20 mm distal to it.

Quantitative assessment was carried out by a point-counting technique. Such a method is easy to use, accurate, and invariant to changes in the number of fibres within a sample area. A square 1 × 1 mm grid was placed on the mounted section and a graticule with a square subdivided into 100 (10 × 10) smaller squares was placed in the eyepiece of the microscope. With the 16 × objective selected, a square of size 1 × 1 mm on the specimen grid corresponded approximately to the whole square in the eyepiece. The square graticule in the eyepiece contained 121 intersections. Every fourth mm^2 throughout a muscle cross-section, referred to in the following as a 'sample area', was selected for measurement. For each of the selected sample areas, the numbers of intersections falling on normal and on degenerating muscle fibres were recorded separately. A fibre was recorded as 'degenerating' when it showed one or more of the following features: a pronounced shift from a polygonal to a rounded shape; hypereosinophilia; invasion of mononuclear cells; hyalinization; vacuolation; disruption of the cytoplasm.

Whereas the transformation of fibre type induced by stimulation has proved consistently reproducible, our own preliminary studies had revealed that this was far from true for stimulation-induced damage. Indeed, such inconsistencies discouraged us from publishing any results until we had the confidence that could come only from a comprehensive study involving larger groups. We recognized that multivariate methods of analysis would be needed to assess the possible influence of experimental technique, variation between individual animals and different muscles, and possible systematic variation in the incidence of damage. By applying such techniques we have been able to consider several variables simultaneously, to investigate interactions between variables, and to apply methods of inferences that allowed us to interpret the results in a comprehensive manner.

Each sample area was characterized by seven variables: *nn*, the number of intersections on normal muscle fibres; *nd*, the number of intersections on degenerating muscle fibres; *id*, the individual rabbit; *exp*, the type of experiment (pattern and frequency of stimulation, operated non-stimulated controls, unoperated controls); *m*, the muscle (TA or EDL); *sec*, the muscle cross-section of which the sample area was part; *area*, the sample area within the muscle cross-section. Two of these variables were combined to give a 'degeneration index', defined as the volume percentage of degenerating muscle fibres estimated from a sample area, $100(nd/(nn + nd))$.

The statistical package SAS (SAS Institute Inc, USA) was used throughout, and the unbalanced analyses of variance (ANOVA) were performed with procedure GLM, which carries out the various tests and forms the necessary residual plots. The response (dependent) variable—the degeneration index—was expressed in terms of the explanatory (independent) variables—*id, exp, m, sec*, and *area*. The effects of the explanatory variables on the response variables were assessed in terms of the significance of the F-statistics. The sources of inhomogeneity, as revealed by the F-statistic, were investigated using a multiple comparison method. For each type of muscle, the degeneration index for each pattern and frequency of stimulation and for the control groups were tested in pairs by a t-statistic.

5.5 Operated non-stimulated and unoperated control groups

The use of two control groups enabled us to separate the effects of stimulation from the sequelae of anaesthesia and surgery associated with the implantation of the stimulators and electrodes. There was no significant difference between the two groups and in neither group was there any more than one degenerating muscle fibre in a sample area comprising several hundred fibres. Thus, any damage resulting from stimulation could not be attributed to operative disturbance *per se* (Lexell *et al.* 1992).

5.6 Variability within and between muscle cross-sections

The extent of degeneration varied considerably between the sample areas, but there was no evidence of systematic variation within a muscle cross-section; the results were not, therefore, influenced by the actual sites selected for sampling within the muscle cross-section (Table 5.1). In the EDL muscle, but not in the TA, there was significantly more degeneration in the proximal portion of the muscle than in the distal portion, especially after continuous stimulation. This variability within and between different levels along the length of a muscle would be large enough to bias the overall results if not taken into account. Therefore, whenever degeneration in a particular muscle is analysed for the first time, several areas should be sampled within any given cross-section, and cross-sections taken from at least two levels within the muscle (Lexell *et al.* 1992).

5.7 Variability between animals

The incidence of degeneration varied markedly between individual rabbits (Lexell *et al.* 1992; Table 5.1). The TA muscles in two rabbits from the group stimulated continuously at 10 Hz, for example, were virtually undamaged, whereas others were more obviously affected. The reasons for this variability are not clear but most probably reflect real interanimal differences in the susceptibility to stimulation-induced damage. This interpretation is supported by the analysis of muscle differences, in which the rabbits that showed most degeneration in the EDL muscle also showed the most degeneration in the TA muscle regardless of the type of stimulation used. Some of the inconsistency in the earlier studies can now be understood as a consequence of the extensive variation between individual animals revealed by this more definitive study.

Table 5.1 The means and standard deviations (SD) of the volume percentage of degenerating muscle fibres per sample area in the tibialis anterior (TA) and extensor digitorum longus (EDL) muscles from 22 stimulated rabbits

Rabbit	TA		EDL	
	Mean	SD	Mean	SD
10 Hz continuous				
1	3.4	1.9	12.8	2.8
2	6.9	4.4	13.9	7.0
3	0.3	0.5	10.5	4.4
4	0.5	0.9	9.3	5.2
10 Hz intermittent				
1	1.5	1.2	5.1	2.2
2	0.9	1.0	3.6	3.7
3	1.2	0.9	4.5	2.2
4	0.5	0.6	2.0	2.2
5	0.8	0.9	5.3	2.2
5 Hz continuous				
1	2.0	1.8	10.4	3.8
2	2.6	3.9	6.7	4.1
3	0.9	1.3	14.2	3.5
4	1.9	1.3	10.1	3.4
2.5 Hz continuous				
1	0.07	0.2	6.8	3.2
2	0.2	0.4	0.3	0.6
3	0.8	1.6	2.4	1.4
4	0.2	0.5	0.1	0.3
1.25 Hz continuous				
1	0.0	0.0	0.4	0.8
2	0.1	0.3	0.0	0.0
3	0.0	0.0	0.03	0.2
4	0.03	0.2	0.07	0.2

5.8   Influence of pattern and frequency

Some investigators have asserted that intermittent patterns of stimulation, such as 8 hours/day, provide a more 'natural' stimulus to muscle than continuous patterns (Pette *et al.* 1973; Seedorf *et al.* 1983); this was

apparently based on undocumented assumptions about the circadian variation of normal motor activity. In later work, those groups adopted a 1 hour
on/1 hour off pattern for many of their experiments. This was one of the
main technical differences between our own procedures and those on which
the reports of extensive damage (Maier *et al.* 1986, 1988) were based, and we
therefore examined it carefully.

If the amount of degeneration were related to the aggregate number of
impulses received by the muscle, one would anticipate that the intermittent
pattern would be less damaging than continuous stimulation at the same
frequency. There is, however, an alternative hypothesis, based on an
ischaemia-reperfusion model, that would predict the opposite result. In
brief, the stimulated muscle is at risk of hypoxia, because the metabolic
demands of continuous activity are accompanied by an increased possibility
of vascular occlusion resulting from a raised mean intramuscular pressure.
Any damage caused by ischaemia during the periods of stimulation could be
amplified by reperfusion with fully oxygenated blood during the periods of
rest, an effect that has been attributed to peroxidation of membrane lipids by
oxygen-derived free radicals (see §§4.4 and 6.2.3).

Our preliminary studies supported the ischaemia-reperfusion hypothesis,
but the larger-scale study yielded a different result: stimulation at 10 Hz
delivered in a 1 hour on/1 hour off pattern was associated with significantly
less damage than 10 Hz continuous stimulation (Lexell *et al.* 1992; Fig. 5.2).
The way in which the incidence of degenerating muscle fibres varied with
the different patterns and frequencies of stimulation shows, however, that
damage is not a simple function of the mean stimulation frequency or the
aggregate number of impulses received by the muscle. For both TA and
EDL muscles, the mean volume percentage of degenerating muscle fibres
declined with declining frequency; at the lowest frequency, 1.25 Hz, both
the TA and the EDL muscles were indistinguishable from unstimulated
control muscles in terms of the numbers of degenerating muscle fibres
(Lexell *et al.* 1993; Fig. 5.2). But the extent of damage for the 10 Hz
intermittent stimulation was also less than for 5 Hz continuous stimulation,
a pattern that delivers the same aggregate number of impulses (Lexell *et al.*
1993; Fig. 5.2). The reasons for this difference are far from clear, but may
provide a clue to the cellular mechanisms underlying the induction of
activity-related damage.

5.9   Variability between TA and EDL muscles

The most striking differences emerged from the analysis of the two muscles
used in the study. There was consistently more damage in the EDL than in
the TA muscle, for every pattern and frequency of stimulation (Lexell *et al.*
1992, 1993; Fig. 5.3). We considered the possibility that the two muscles were

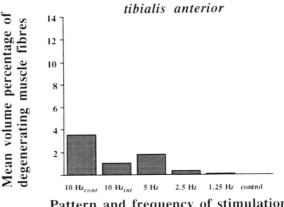

Pattern and frequency of stimulation

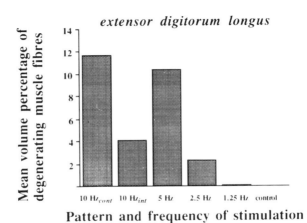

Pattern and frequency of stimulation

Fig. 5.2 The effects of different patterns and frequencies of stimulation on tibialis anterior and extensor digitorum longus muscles.

responding asynchronously to stimulation, so that degenerative processes were more advanced in the EDL muscle at the time of sampling. However, the greater susceptibility to damage of the EDL muscles was a finding that had also emerged powerfully and consistently from other experiments in this laboratory, including the preliminary studies, in which the muscles were sampled at other times. The most likely interpretation is that the two muscles differ in their susceptibility to stimulation-induced damage. This observation suggests a plausible explanation for the apparent discordance between the observations reported by Maier *et al.* (1986, 1988) and those of other groups: Maier and his colleagues based their conclusions about the role of degenerative and regenerative phenomena exclusively on data from the EDL muscle, whereas the conflicting evidence had come largely from experiments carried out with the TA muscle.

This differential susceptibility of the TA and EDL muscles to stimulation is not easy to explain but it appears to be quite real, and has also been observed by Fridén and Lieber in the response of the muscles to cyclic eccentric contraction (see §3.5). From an anatomical viewpoint, the muscles and their nerve and blood supplies are very closely associated. Their fibre type compositions are broadly similar (Lexell *et al.* 1994). Both muscles are physiological dorsiflexors of the ankle, but the EDL muscle acts across two joints and could therefore be stretched differently to the TA muscle during postural movements of the animal. The fibre architecture of the two muscles does differ, the TA muscle being essentially parallel-fibred, and the EDL pennate. It remains to be established whether kinesiological considerations such as these or differences at the cellular level are responsible for the disparate behaviour of the two muscles. At all events, it is clearly a mistake to conclude that a stimulation regime that causes damage to one muscle will necessarily induce damage to a similar extent in all other muscles.

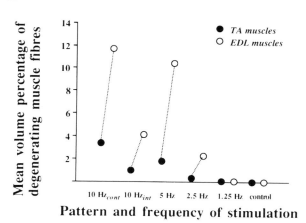

Fig. 5.3 The difference between tibialis anterior (TA) and extensor digitorum longus (EDL) muscles in their response to different patterns and frequencies of stimulation.

5.10 Damage and type transformation

In the TA muscle, type 1 fibres normally constitute 3–5 per cent of the total (Lexell *et al.* 1994). Fast-to-slow type transformation in this muscle therefore involves at least 95 per cent of the fibres. Against this figure, the mean volume percentage of less than 3.4 per cent of degenerating fibres recorded in the TA muscle for 10 Hz continuous stimulation is insignificant. Turnover of fibres might contribute more significantly to fibre type conversion in this muscle if it were ongoing, but the evidence is against this. According to Maier *et al.* (1986), the process reaches its maximum development between 8 and 10 days from the onset of stimulation. Our own observations, on muscles stimulated from a few hours to 12 weeks, confirm that damaged fibres are very rare after 3 weeks of stimulation. Thus, type transformation is a distinct phenomenon and any contribution from damage brought about by stimulation is incidental, not obligatory (Lexell *et al.* 1992). In more extensively damaged muscle this may be a significant contributor to the overall fibre type transformation. The evidence of neonatal myosin reported by Maier *et al.* (1988) is therefore consistent with the degree of degeneration and regeneration they recorded in the EDL muscle, but does not conflict with the earlier conclusion (W.E. Brown *et al.* 1983; Hoffman *et al.* 1985) that transformation does not require recapitulation of a developmental sequence.

5.11   Other potential sources of damage

The chronic stimulation technique that has been developed, and progressively refined, in this laboratory depends on the use of small loop electrodes, one of which is placed very close to the common peroneal nerve. This confines the stimulating current field, enabling us to achieve supramaximal stimulation with a low stimulating voltage. Technical practices in other laboratories differ, and it is worth pointing out that high stimulating voltages in the hindlimb, sufficient to cause some spread of stimulation to the tibial nerve, would recruit the powerful plantar flexors of the ankle. Under these conditions, the muscles of the anterior compartment would be simultaneously stretched and stimulated, with a potential for damage similar to that caused by the performance of eccentric work.

5.12   Conclusions and clinical implications

Supramaximal electrical stimulation of a motor nerve is an artificial means of activating a muscle, resulting as it does in synchronous activation of all motor units. The remarkable feature of muscle is that it accepts the new burden of activity, and adapts to it, with so little damage. Some damage is, nevertheless, sustained by the stimulated muscle, and the present study illustrates the complexity of the phenomena involved.

An operational problem is that observations have to be made in the presence of marked intra- and inter-animal variation: this must be properly accommodated in a rigorous experimental design. When this has been done, certain clear conclusions emerge. Different anatomical muscles are differentially susceptible to the degenerative process, and observations confined to a specific muscle should not, therefore, be regarded as illustrative of a general principle. The fact that stimulated-induced transformation of muscle type can take place in a muscle that has sustained an insignificant level of damage shows that transformation is distinct from damage: both phenomena may take place in response to stimulation, but one is not responsible for the other. Most importantly, damage is not an inevitable consequence of any type of electrical stimulation, but is related closely to the pattern of activation.

The extent of degeneration is not related in a simple way to total activity, for an intermittent pattern was found to be less damaging than a continuous pattern that delivered the same aggregate number of impulses. We would expect this effect to vary with the actual on/off times, as well as the duty cycle: when the on/off periods are short—of the order of a few seconds each—the extent of damage should be similar to that of continuous stimulation at the equivalent mean frequency, whereas differences would be predicted to emerge as the on/off periods extended into minutes.

The mechanisms that lead to stimulation-induced damage could be triggered either by events intrinsic to the cell—for example, the metabolic challenge posed by increased activity—or by extrinsic events—for example, mechanical stretch or vascular occlusive effects. The overall susceptibility of a muscle to damage may depend on a critical combination of these factors. This might explain some of the variability we have observed in the response to a given regime—such as the intermittent pattern, which was found to be less damaging than continuous stimulation in the present study, even though it was more damaging in the earlier experiments. It might also account for some of the variation between individual animals and between specific muscles.

On a cautionary note, degenerative events have a time course similar to those of regulatory events, changes in nucleic acid synthesis, and early changes in muscle-type-specific proteins associated with stimulation-induced transformation. It would therefore be wise to check that observations attributed to type transformation are not due to degenerative-regenerative phenomena. The present study shows that results that are obtained, or at least confirmed, in the TA muscle of the rabbit are likely to be free of these ambiguities.

The prospects for the surgical application of transformed skeletal muscle grafts have prompted experimental work that involves a variety of patterns of stimulation, muscles, and species. Just the process of mobilizing a pedicle graft incurs some damage as a result of ischaemia, tenotomy, and mechanical trauma. The work reviewed here shows that it is important to evaluate the additional contribution that results from stimulation, and this issue will be addressed in more detail in Chapter 11. A better understanding of the interaction between all these factors would provide a systematic route to the design of escalating stimulation regimes for clinical use that produce the desired properties in the long term without compromising the viability of the muscle fibres in the short term.

Acknowledgements

The studies were carried out while Jan Lexell was working during 1989–90 at the University of Liverpool, England, supported by grants and scholarships from the Swedish Work Environment Fund, the Swedish Institute, the Swedish Society of Medicine, the Research Council of the Swedish Sports Federation, the Tore Nilsson Foundation, and the Hans and Loo Osterman Foundation. The authors also acknowledge the support given to this research by The British Heart Foundation. Mrs J. Currie and Ms H. Sutherland provided technical assistance. Dr C.N. Mayne contributed valuable help in setting up the experimental aspects of the study, and the unpublished work of Miss E. Robinson, Dr J.P. Cooper, and Miss J. Williams helped to define in advance many of the problems that were addressed in greater detail here.

References

Brown, J.M.C., Henriksson, J., and Salmons, S. (1989). Restoration of fast muscle characteristics following cessation of chronic stimulation in the rabbit: physiological, histochemical and metabolic changes during slow-to-fast transformation. *Proceedings of the Royal Society of London (Series B)*, **235**, 321–46.

Brown, W.E., Salmons, S., and Whalen, R.G. (1983). The sequential replacement of myosin subunit isoforms during muscle type transformation induced by long term electrical stimulation. *Journal of Biological Chemistry*, **258**, 14686–92.

Buller, A.J., Eccles, J.C., and Eccles, R.M. (1960). Differentiation of fast and slow muscles in the cat hind limb. *Journal of Physiology (London)*, **150**, 399–416.

Eisenberg, B.R. and Salmons, S. (1981). The reorganisation of subcellular structure in muscle undergoing fast to slow type transformation. *Cell and Tissue Research*, **220**, 449–71.

Eisenberg, B.R., Brown, J.M.C., and Salmons, S. (1984). Restoration of fast muscle characteristics following cessation of chronic stimulation. *Cell and Tissue Research*, **238**, 221–30.

Hoffman, R.K., Gambke, B., Stephenson, L.W., and Rubinstein, N.A. (1985). Myosin transitions in chronic stimulation do not involve embryonic isozymes. *Muscle and Nerve*, **8**, 796–805.

Jarvis, J.C. and Salmons, S. (1991). A family of neuromuscular stimulators with optical transcutaneous control. *Journal of Medical and Engineering Technology*, **15**, 53–7.

Lexell, J., Jarvis, J., Downham, D., and Salmons, S. (1992). Quantitative morphology of stimulation-induced damage in rabbit fast-twitch skeletal muscles. *Cell and Tissue Research*, **269**, 195–204.

Lexell, J., Jarvis, J., Downham, D., and Salmons, S. (1993). Stimulation-induced damage in rabbit fast-twitch skeletal muscles: a quantitative morphological study of the influence of pattern and frequency. *Cell and Tissue Research*, **273**, 357–62.

Lexell, J., Jarvis, J.J., Downham, D.Y., and Salmons, S. (1994). The fibre type composition of rabbit tibialis anterior and extensor digitorum longus muscles. *Journal of Anatomy*, **185**, 95–101.

Maier, A., Gambke, B., and Pette, D. (1986). Degeneration-regeneration as a mechanism contributing to the fast to slow conversion of chronically stimulated fast-twitch rabbit muscle. *Cell and Tissue Research*, **244**, 635–43.

Maier, A., Gorza, L., Schiaffino, S., and Pette, D. (1988). A combined histochemical and immunohistochemical study on the dynamics of fast-to-slow fiber transformation in chronically stimulated rabbit muscle. *Cell and Tissue Research*, **254**, 59–68.

Pette, D. and Schnez, U. (1977). Coexistence of fast and slow type myosin light chains in single musclefibres during transformation as induced by long term stimulation. *FEBS Letters*, **83**, 128–30.

Pette, D., Smith, M.E., Staudte, H.W., and Vrbová, G. (1973). Effects of long-term electrical stimulation on some contractile and metabolic characteristics of fast rabbit muscle. *Pflügers Archiv*, **338**, 257–72.

Rubinstein, N., Mabuchi, K., Pepe, F., Salmons, S., Gergely, J., and Sréter, F. (1978). Use of type-specific antimyosins to demonstrate the transformation of individual fibers in chronically stimulated rabbit fast muscles. *Journal of Cell Biology*, **79**, 252–61.

Salmons, S. (1967). An implantable electrical stimulator. *Journal of Physiology (London)*, **188**, 13–14P.

Salmons, S. and Henriksson, J. (1981). The adaptive response of skeletal muscle to increased use. *Muscle and Nerve*, **4**, 94–105.

Salmons, S. and Jarvis, J.C. (1991). A simple optical switch for implantable devices. *Medical and Biological Engineering and Computing*, **29**, 554–6.

Salmons, S. and Sréter, F. (1976). Significance of impulse activity in the transformation of skeletal muscle type. *Nature (London)*, **263**, 30–4.

Salmons, S. and Vrbová, G. (1969). The influence of activity on some contractile characteristics of mammalian fast and slow muscles. *Journal of Physiology (London)*, **201**, 535–49.

Seedorf, K., Seedorf, U., and Pette, D. (1983). Coordinate expression of alkali and DTNB myosin light chains during transformation of rabbit fast muscle by chronic stimulation. *FEBS Letters*, **158**, 321–4.

6

Intracellular mechanisms involved in skeletal muscle damage

Anne McArdle and Malcolm J. Jackson

6.1 Introduction

It is widely believed that cells subjected to a damaging or lethal degree of stress pass through a number of different stages and that damage ultimately occurs through one of a relatively few common final pathways. Of these, the ones usually considered are loss of energy supply to the cell, loss of intracellular calcium homeostasis, and overactivity of oxidizing, free radical-mediated reactions. These postulated pathways have been invoked as part of the damaging process in a variety of cell types, and all have been implicated in the process by which muscle cell damage occurs after different stresses (De Leiris and Feuvray 1979; Davies *et al.* 1982; Jackson *et al.* 1991*a*).

The aim of this chapter is to examine the mechanisms by which muscle cells that are subjected to stress may be damaged, the ways in which these processes target specific cell components, and the sites at which they interact. We will end by outlining the problems involved in attempting to attribute muscle damage in any particular situation to one specific mechanism.

6.2 Major intracellular mechanisms involved in muscle damage

6.2.1 Calcium in skeletal muscle damage

Intracellular calcium accumulation has long been implicated in cell death (Schanne *et al.* 1979), and there is a considerable literature on the role of calcium in muscle damage (Jackson *et al.* 1991*a*).

Early studies by Duncan (1978) demonstrated that an increase in intracellular calcium content (induced by treatment with the calcium ionophore, A23187) caused damage to the myofilaments of skeletal muscle. Duncan and his coworkers (Duncan 1978; Publicover *et al.* 1978) proposed that the ionophore acted by releasing calcium from the sarcoplasmic reticulum (SR), and the idea that a calcium-induced calcium release was responsible received support from the studies of Publicover *et al.* (1978) and Soybell *et al.* (1978). Jones *et al.* (1984) showed that efflux of cytosolic enzymes and ultrastructural damage caused by excess contractile activity was reduced when extracellular calcium was removed, and they proposed that the damaging process was initiated by extracellular calcium accumulated during the activity.

Anand and Emery (1980) reported that the plasma membrane calcium channel blocker, verapamil, inhibited the calcium-stimulated efflux of cytosolic proteins from incubated strips of normal human muscle. However, Jones *et al.* (1984), working with an isolated mouse muscle model, could not show any protective effect of calcium channel blockers such as verapamil and nifedipine against contraction-induced damage. This study suggested that elevation of intracellular calcium under these conditions could not be attributed solely to classical calcium channels.

There is therefore evidence that abnormalities of calcium homeostasis are involved in damage to normal skeletal muscle induced by 'physiological' stresses, and this extends to disease processes such as Duchenne muscular dystrophy (King Engel 1978; Duncan 1978) and malignant hyperthermia (Arthur and Duthie 1991; Duthie and Arthur 1993).

6.2.2 Mechanisms by which calcium accumulation causes damage to muscle

Several different processes have been proposed to explain how muscle could be damaged by an elevated intracellular calcium content. These include:
- stimulation of calcium-activated proteases;
- activation of lysosomal processes;
- mitochondrial overload;
- activation of lipolytic enzymes.

These will be considered in turn.

6.2.2.1 Calcium activation of proteases

Calcium-activated proteases are non-lysosomal intracellular enzymes that are found in many cells. They are dependent on elevated intracellular levels of calcium for activation, but they do not require ATP. DeMartino and Croall (1987) described two distinct calcium-dependent proteases, calpain type I (or CDP I) and calpain type II (or CDP II), that are activated at different levels of calcium (micromolar and millimolar, respectively).

Tissues that contain these proteases also contain an inhibitor protein (in some cases calpastatin), and an activator protein. Little is known about these proteins, but protease activity depends on the balance between them, and caution has to be taken if protease activities in cells are measured without measuring the related peptides at the same time. It has been proposed that the inhibitors protect the cell membrane against damage during transient increases in the level of free calcium (Mellgren 1987).

Calcium-dependent proteases may be involved in general protein degradation in several cell types. The proteases are not specific for particular protein sequences, but may still produce specific damage within the muscle because of compartmentation. They have been associated with Z-disc streaming (see §3.1 and Fig. 3.2) and A-band disruption (Cullen and Fulthorpe 1982). Inhibitors of the proteases, such as leupeptin, did not prevent the appearance of some features of muscle damage induced by the calcium ionophore (Duncan *et al.* 1979, 1980; Jackson *et al.* 1984*a*), but the lack of effect may be due to an inability to cross the lipid membrane to reach the necessary sites of action.

6.2.2.2   *Calcium activation of lysosomal proteases*

Rodemann and coworkers (Rodemann and Goldberg 1982; Rodemann *et al.* 1982) found evidence that lysosomal enzymes were activated during calcium ionophore-induced damage to muscle. They suggested that an ionophore-induced increase in intracellular calcium caused an increase in prostaglandin E_2 production that stimulated lysosomal activity. However, other workers have been unable to reproduce these results (Baracos *et al.* 1986), and experiments by Duncan *et al.* (1979), involving the use of lysosomal inhibitors, indicated that lysosomal protease activity was not a major factor in muscle damage caused by the action of the ionophore.

6.2.2.3   *Mitochondrial calcium overload*

In extreme cases of elevated intracellular calcium, mitochondria will accumulate significant amounts of calcium (up to 3 µmol calcium/mg mitochondrial protein) accompanied by phosphate, causing calcium phosphates to precipitate in the mitochondria (Gohil *et al.* 1988). Increases in mitochondrial calcium content in this range have been shown to inhibit mitochondrial respiration (Wrogemann and Pena 1976). Thus mitochondria may try to maintain calcium homeostasis at the expense of energy production. When they are no longer able to cope with increasing levels of intracellular calcium, the mitochondria become overloaded and release their calcium stores into the cytoplasm.

Wrogemann and Pena (1976) proposed that mitochondrial overload with calcium was a universal precursor to skeletal muscle damage. However, Jackson *et al.* (1984*a*) were able to show that muscle damage could be brought about by an external calcium-dependent mechanism even when mitochondrial poisons had been used to inhibit mitochondrial activity, implying that mitochondrial overload was an accompaniment, rather than a prerequisite, of muscle damage.

6.2.2.4   *Calcium activation of lipases*

Calcium activation of phospholipases has been implicated in muscle damage (Rodemann *et al.* 1982). Jackson *et al.* (1984*b*) suggested that elevated intracellular levels of calcium caused damage to muscle by activating phospholipases, which utilize membrane phospholipids as substrates. The resulting lysophospholipids and free fatty acids disrupt the membrane because of their detergent-like properties (Jackson *et al.* 1991*a*). This theory was supported by work on cardiac tissue by Das *et al.* (1985), who further proposed that reacylation of lysophospholipids is inhibited during damage, enhancing the deleterious process already described. However, Duncan and Jackson (1987) found that non-specific phospholipase inhibitors do not inhibit the damage process completely. They showed that the calcium ionophore, A23187, caused efflux of the intracellular enzyme creatine kinase (CK) by a system that was blocked by phospholipase inhibitors, whereas structural damage to the muscle was unaffected by these substances.

The prostaglandins, which are byproducts of phospholipase A_2 activity, have themselves been implicated in muscle damage. Prostaglandins may be responsible for a number of the features of muscle damage induced by calcium accumulation (Rodemann *et al.* 1982).

It is clear from the preceding discussion that no one process is responsible for the damage to skeletal muscle that occurs following a rise in intracellular calcium. Rather it appears that when calcium levels rise above a certain level it is a combination of catastrophic processes that leads to the death of the cell.

6.2.3   Free radicals in skeletal muscle damage

It is recognized that there are a number of potential sites for the production of free radicals within muscle, such as the mitochondrial electron transport system (Boveris and Chance 1973), membrane-bound oxidases (Crane *et al.* 1985), and infiltrating phagocytic cells (Font *et al.* 1977). In addition, xanthine oxidase, which is located within endothelial tissue in muscle, is a potential site for free-radical production (Karthuis *et al.* 1985; McCord 1985). Muscle is unique in its ability to undergo very rapid, co-ordinated changes in energy supply under conditions of repeated contraction. Since the changes require major variations in the oxygen flux through the tissue and the electron flux through the mitochondrial respiratory chain, they may predispose to the formation of oxygen-centred free-radical species (Boveris *et al.* 1972).

Free radicals can cause damage by lipid peroxidation of unsaturated fatty acids in skeletal muscle membranes (Davies *et al.* 1982). They can also cause oxidative damage to DNA and proteins, including the protective anti-oxidant enzymes (Halliwell and Gutteridge 1985). Both the secondary and tertiary structures of proteins depend on the number and location of free sulfhydryl groups and intramolecular disulfide bonds. The oxidation of adjacent thiol groups to form intramolecular disulfides, or the formation of intermolecular mixed thiols by interaction of a protein with a low molecular weight thiol compound such as glutathione, can modify protein structure and thus influence enzyme activity. Both the Na^+/K^+-ATPase of the plasma membrane (Harris and Stahl 1980) and the Ca^{2+}-ATPase of the SR (Scherer and Deamer 1986) contain sulfhydryl groups that can be oxidized in this way, resulting in a loss of activity. These protein thiolation processes are known to occur at a very early stage of oxidative stress (Thomas *et al.* 1994).

Free radicals have been implicated in the muscle degeneration that takes place in a number of disorders, including Duchenne muscular dystrophy (Murphy and Kehrer 1986) and malignant hyperthermia (Arthur and Duthie 1991; Duthie and Arthur 1993). There has also been considerable speculation that free radicals are involved in exercise-induced muscle damage. Davies *et al.* (1982) reported increased production of free-radical species, detected by electron spin resonance (ESR) techniques, following extensive exercise in rats. Vitamin E provides protection against free-radical damage by donating hydrogen atoms to free radicals, and can thus inhibit a variety of degradative reactions associated with lipid peroxidation (Brady *et al.* 1978; Gee and Tappel 1981). Rats deficient in vitamin E had a decreased endurance capacity and showed increased free-radical ESR signals; the present authors therefore proposed that the decreased endurance capacity was due to free radical-mediated peroxidative damage. Initial studies by Jackson *et al.* (1985), also using ESR, supported an involvement of free radicals in skeletal muscle damage, but further work has shown that the ESR signal observed is not related directly to the process of cellular damage (Johnson *et al.* 1988).

Cells have various mechanisms for protecting themselves from free-radical damage, including free-radical scavengers and anti-oxidant enzymes. It has been shown that these mechanisms are upregulated in athletes (Robertson *et al.* 1991) presumably as a protection against exercise-induced increases in free-radical production. This idea was consistent with the work of Zerba *et al.* (1990) who showed that superoxide dismutase had a protective effect, reducing delayed onset damage in mouse skeletal muscle. Jackson *et al.* (1991*b*) studied the release of glutathione from isolated skeletal muscles. Glutathione is a substrate for glutathione peroxidase, an enzyme that acts to reduce the toxic effects of oxygen free radicals. Some cells actively remove oxidized glutathione from the cytosol (Chance *et al.* 1979); the authors therefore proposed that increased glutathione efflux reflected an increase in free-radical activity within skeletal muscle, as has been reported for cardiac muscle (Curello *et al.* 1987). Packer (1984) and Gohil *et al.* (1988) had previously interpreted glutathione oxidation and release in muscular exercise as an index of free-radical activity. Jackson *et al.* (1991*b*) confirmed that glutathione, like CK, was indeed released from damaged skeletal muscle, but release did not appear to be a prerequisite of damage.

A further endogenous cellular defence mechanism has emerged in recent years. All cells respond to a short-term sublethal elevation in temperature by synthesizing proteins known as 'heat-shock' or 'stress' proteins. These appear to play an important role in the thermotolerance that these cells then show (Gerner and Schneider 1975; Li and Werb 1982; Riabowol *et al.* 1988). Recent evidence suggests that this same group of proteins may be cytoprotective against other lethal stresses. Particularly important in this regard is the induction of such proteins by oxidative stress and the apparent protection that these proteins confer against a subsequent exposure to oxidative stress (Polla *et al.* 1991). Several recent studies have demonstrated that hyperthermia-induced stress proteins can protect cardiac tissue against damage induced by a period of ischaemia and reperfusion (Currie *et al.* 1988; Donnely *et al.* 1992; Yellon *et al.* 1992; Marber 1994). Similarly, Garramone and his colleagues (1994) have shown that prior heat stress provides significant biochemical and ultrastructural protection against ischaemic injury in rat skeletal muscles.

The role of free radicals in skeletal muscle damage is confusing and incompletely understood. This may be due to the different experimental models in use, but it probably also reflects the paucity of definitive methods for the detection and measurement of free-radical species in tissues.

6.2.4 Decreased cellular energy supply

It has long been recognized that a fall in cellular ATP content is associated with cell death and that muscle ATP levels can fall in response to stresses of different types. De Leiris and Feuvray (1979) have suggested that release of cellular proteins occurs when cellular ATP falls below a certain critical level, and Pennington (1988) proposed that interference in the energy supply to the muscle membrane was an important factor leading to enzyme efflux.

The idea that energy depletion causes an increase in plasma membrane permeability was supported by the work of Cerny and Haralambie (1983) on human subjects who undertook excessive exercise; their results indicated that the intensity of the exercise was the primary factor responsible for elevated levels of circulating muscle-derived enzymes. These authors postulated that the energy demanded by working muscles reduced the amount of ATP available for maintenance of membrane integrity, leading eventually to a loss of cellular enzymes and damage.

Several reports in the literature have pointed to a fall in energy levels as the primary cause of enzyme efflux from muscles. However, most authors agree that there is a significant delay between reduction in energy and loss of enzymes, and that the fall in intracellular energy probably initiates another mechanism that leads ultimately to loss of cytosolic components. Jones *et al.* (1983) examined the factors influencing enzyme efflux from isolated skeletal muscles subjected to excessive contractile activity and suggested that a failure of muscle energy supply was a key step in the damaging process.

The idea is supported by data from cardiac tissue. In a review of the effects of ischaemia on cardiac muscle, Hearse (1979) suggested that loss of membrane control may arise from an energy shortage during the early stages of cardiac ischaemia or anoxia; a progressive fall in energy reserves is generally observed, but damage to the cell membrane is not apparent until much later. Michell and Coleman (1979) proposed that the permeability changes during cardiac ischaemia are initiated by homeostatic problems, which arise either directly or indirectly from energy depletion and progress only later to become overt physical membrane lesions. In instances of regional cardiac ischaemia, substantial depletion of ATP occurred in the first 120 min, at a time when tissue CK content had not decreased (Gillis 1985). A study of isolated mammalian cardiomyocytes during prolonged anoxia led Allshire *et al.* (1987) to suggest that the fall in ATP content induced by anoxic conditions led to the onset of rigor, and that rigor-induced contracture activated myosin ATPase, which further depleted ATP. Eventually the ATP level would be too low to drive the sarcolemmal and SR calcium pumps, and the result would be impaired calcium homeostasis.

Although there appears to be a substantial association between a fall in muscle ATP content and muscle damage, there are exceptions: for example, West-Jordan *et al.* (1990, 1991) demonstrated damage to skeletal muscle *in vitro* that was not associated with a fall in cellular ATP content.

6.3   Interrelationships between mechanisms underlying damage

We have considered three intracellular mechanisms associated with muscle damage: calcium overload; free radical formation; and a decrease in energy supply. It is worth considering whether these mechanisms operate in isolation in situations where muscle damage occurs. A number of studies indicate that this is not the case and examples of data suggestive of significant interaction between the pathways will be considered below.

Overt damage to muscle—for example, by direct trauma—will lead to a loss of cell viability with a consequent failure of cells to maintain transmembrane gradients of ions such as calcium, a loss of cellular energy supplies, and stimulation of free radical-mediated degenerative processes, such as lipid peroxidation (Halliwell and Gutteridge 1984). There are other circumstances in which evidence can be found for disturbances in all three main pathways.

6.3.1 Effects of failure of muscle energy supply on calcium homeostasis and free-radical production

A failure of muscle energy supply (or a severe reduction in ATP content) can lead to a failure of muscle calcium homeostasis in several ways. The SR and sarcolemmal calcium pumps of skeletal muscle are both ATP-dependent (Schatzmann 1975); thus under low-energy status these pumps will not function, resulting in a gradual accumulation of cytosolic calcium. A direct effect of a lack of ATP on membrane composition (due, for example, to defective ATP-dependent reacylation of lysophospholipids) could lead to an influx of external calcium driven by the large extracellular-to-intracellular electrochemical gradient for this element. There is considerable evidence that such a mechanism is involved in damage to skeletal muscle under experimental conditions. Jackson *et al.* (1984*b*) and Duncan and Jackson (1987) reported that treatment of muscle with inhibitors of mitochondrial activity caused damage to muscle that could be inhibited partially by removal of external calcium, and West-Jordan *et al.* (1990) found that mitochondrial inhibitors caused a very rapid loss of cellular ATP content that was associated with a slower influx of external calcium and activation of calcium-dependent degenerative processes (see Table 6.1). Similar processes seem to be active in cardiac tissue subjected to a period of ischaemia and reperfusion (Hearse 1979).

Table 6.1 The effect of treatment of skeletal muscle with 2,4 dinitrophenol (DNP) in the presence or absence of extracellular calcium*

	ATP/P_i ratio	LDH release (mU/30 min)
Prior to DNP treatment ($+Ca^{2+}$)	2.6 ± 0.2	26 ± 3
30 min post-DNP treatment ($+Ca^{2+}$)	0	210 ± 48
Prior to DNP treatment ($-Ca^{2+}$)	2.6 ± 0.3	24 ± 8
30 min post-DNP treatment ($-Ca^{2+}$)	0	62 ± 10

* Data derived from West-Jordan *et al.* (1990) and Jackson *et al.* (1984*a*). Isolated muscles were treated with 2,4 dinitrophenol (DNP; 200 μM) for 30 min in the presence ($+Ca^{2+}$) or absence ($-Ca^{2+}$) of external calcium. Muscle ATP/P_i ratios were measured by ^{31}P-nuclear magnetic resonance spectroscopy.
LDH, Lactate dehydrogenase.

In recent years there has been considerable speculation about possible links between failure of energy production and the generation of free-radical species. This would appear to be particularly important in damage induced by ischaemia and subsequent reperfusion. Most of the data available concern the role of xanthine oxidase as a system for generating superoxide radicals (McCord 1985). Breakdown of ATP furnishes hypoxanthine as a substrate for xanthine oxidase, a form of xanthine dehydrogenase modified by a calcium-dependent protease. Xanthine dehydrogenase may not exist in skeletal muscle, but it is present in endothelial cells within the muscle bulk, and this provides a potential site for this mechanism.

6.3.2   Effects of increased free radical activity on calcium homeostasis and muscle energy supply

Since free-radical species can damage all biomolecules it is possible to envisage a number of ways in which free radicals could disturb calcium homeostasis. There is specific evidence for inactivation of plasma membrane pumps responsible for extrusion of calcium from cells (Nicotera *et al.* 1985) and for free radical-mediated damage to intracellular organelles that compromises their ability to sequester excess calcium from the cytosol (Bellomo *et al.* 1985). Both mechanisms could lead to an inability to maintain cellular calcium homeostasis. Once calcium homeostasis is lost then the rise in intracellular calcium will activate the contractile apparatus and energy supplies will be rapidly consumed. Free radical-mediated peroxidation of membrane lipids could also damage mitochondria, with deleterious consequences for ATP production and cellular $NAD(P)H^+$ content.

6.3.3   Effects of a loss of calcium homeostasis on free-radical activity and muscle energy supply

The effects of an increased intracellular calcium content on the ability of mitochondria to generate ATP has been discussed previously, but there is growing evidence that a loss of calcium homeostasis can also influence free-radical reactions. Thus increased intracellular calcium levels activate phospholipase enzymes (Jackson *et al.* 1984*b*), which release free fatty acids from membranes and appear to render them more susceptible to free radical-mediated lipid peroxidation (Cotterill *et al.* 1990). It also provides the free arachidonic acid for production of prostanoids—many of which are free radicals—and a calcium-dependent protease can convert xanthine dehydrogenase to xanthine oxidase, whose action generates superoxide radicals.

In an attempt to examine these phenomena at an experimental level we have recently studied the effect of calcium accumulation on vitamin

E-deficient muscle. Vitamin E is thought to be the primary lipid-soluble scavenger of free radicals in the body, and its absence is known to increase the sensitivity of muscle to contraction-induced damage (Jackson and Edwards 1988). It has been assumed that the primary damaging process in such muscles is enhanced generation of free-radical species during contractile activity (Davies *et al.* 1982). We have examined calcium influx during excessive contractile activity in vitamin E-deficient muscle, damage being assessed by release of CK activity. In earlier studies we had demonstrated a direct correlation between CK efflux from, and calcium accumulation by, muscles (McArdle *et al.* 1991, 1992).

We predicted that excessive contractile activity would lead to increased free-radical activity and hence a greater influx of calcium and release of CK activity from vitamin E-deficient than from control muscles. However, this was not the case (Fig. 6.1). Calcium accumulation by vitamin E-deficient muscle was similar to that of control muscle, yet release of CK was greater than control. These results would indicate that vitamin E-deficient muscle has an increased susceptibility to the deleterious effects of calcium accumulation, and this may be mediated by calcium-activation of free radical-mediated processes, rather than by contractile activity *per se.*

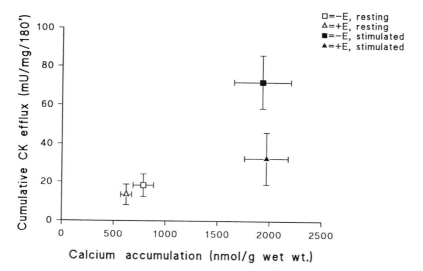

Fig. 6.1 Comparison between the release of creatine kinase activity and the influx of calcium by vitamin E-deficient (-E) and vitamin E-supplemented (+ E) muscles. The methods used have been described in McArdle *et al.* (1992) and Phoenix *et al.* (1990). (Data from McArdle *et al.* 1993.)

6.3.4   Implications for studies of skeletal muscle damage

It is clear from the preceding discussion that all three major pathways are involved in any substantial muscle damage and that it will be difficult to identify primary mechanisms. For example, Z-disc streaming is thought to be an indicator of the activity of calcium-dependent proteases (Cullen and Fulthorpe 1982), but it is conceivable that where such changes are observed they may be secondary to an increase in free-radical activity. Moreover, different aspects of the damage to muscle cells may appear to be mediated by different pathways when the primary cause is in fact the same. Thus Duncan and Jackson (1987) reported that enzyme efflux following treatment of muscle with the mitochondrial inhibitor DNP was dependent on extracellular calcium but myofilament damage was not. Such apparent discrepancies could be explained by a primary lack of ATP; this would lead to both loss of calcium homeostasis and enhancement of free-radical reactions, which would then mediate different aspects of the damaging process.

A further complication is that pharmacological agents can have multiple actions. For example, many agents that inhibit calcium-activated phospholipases have also been shown to be inhibitors of lipid peroxidation, probably through a non-specific effect on membranes (Jackson *et al.* 1984*b*).

6.4   Programmed cell death in skeletal muscle

It is commonly assumed that programmed cell death does not occur in skeletal muscle and therefore has no place in a discussion such as this. There are, however, a number of situations in nature where muscles are lost by programmed cell death. Notable examples are the loss of the muscles of the tadpole tail during metamorphosis and the loss of the intersegmental muscles of the tobacco hawkmoth (*Manduca sexta*), both of which have been studied extensively (Schwartz *et al.* 1990, 1993; Schwartz and Osborne 1993). It is interesting that cell death in these situations does not show the features typical of apoptosis, and that despite its programmed nature many of the morphological features would more usually be associated with processes leading to necrosis (Schwartz *et al.* 1990). However, in a recent paper, several classical features of apoptosis were identified in degenerating skeletal muscle of the *mdx* dystrophic mouse (Tidball *et al.* 1995).

There is evidence from other systems that programmed cell death may involve a failure of calcium homeostasis (Lam *et al.* 1994) and/or increased oxygen radical activity (Hockenbery *et al.* 1993); hence this is another potential route by which these mechanisms may be initiated in skeletal muscle.

6.5 Conclusions

Attempts to ascribe any form of damage to a single intracellular mechanism are complicated by the significant interactions that occur between the major pathways. When tissue damage is advanced, a complex series of biochemical events is set in train that contributes to different facets of the damage process. Pharmacological intervention is likely to be effective only where the primary damaging events can be identified and targeted.

Acknowledgements

The authors would like to thank the many coworkers who have contributed to the work described here. Financial support from The Wellcome Trust, Linbury Trust, Muscular Dystrophy Group of Great Britain and Northern Ireland, and F. Hoffman-La Roche Ltd is gratefully acknowledged.

References

Allshire, A., Piper, H.M., Cuthbertson, K.S.R., and Cobbold, P.H. (1987). Cytosolic free Ca^{2+} in single heart cells during anoxia and reoxygenation. *Biochemical Journal*, **244**, 381–5.

Anand, R. and Emery, A.E.H. (1980). Verapamil and calcium-stimulated enzyme efflux from human skeletal muscle. *Research Communications in Chemical Pathology and Pharmacology*, **244**, 541–50.

Arthur, J.R. and Duthie, G.G. (1991). Malignant hyperthermia: the roles of free radicals and calcium. In *Calcium, oxygen radicals and cell damage* (ed. C.J. Duncan), pp. 115–38. Cambridge University Press.

Baracos, V., Greenberg, R.E., and Goldbert, A.L. (1986). Influence of calcium and other divalent cations on protein turnover in rat skeletal muscle. *American Journal of Physiology*, **250**, E702–10.

Bellomo, G., Richelmi, P., Mirabelli, F., Marinoni, V., and Abbagano, A. (1985). Inhibition of liver microsomal calcium ion sequestration by oxidative stress: role of protein sulphydryl groups. In *Free radicals in liver injury* (ed. G. Poli, K. H. Cheeseman, M.U. Dianzani, and T.F. Slater), pp. 139–42. IRL Press, Oxford.

Boveris, A. and Chance, B. (1973). The mitochondrial generation of hydrogen peroxide. *Biochemical Journal*, **134**, 707–16.

Boveris, A., Oshino, N., and Chance, B. (1972). Cellular production of hydrogen peroxide. *Biochemical Journal*, **128**, 617–30.

Brady, P.S., Brady, L.J., and Ulrey, D.E. (1978). Selenium, vitamin E and the response to swimming stress in the rat. *Journal of Nutrition*, **109**, 1103–9.

Cerny, F.J. and Haralambie, G. (1983). Exercise-induced loss of muscle enzymes. In *Biochemistry of exercise* (ed. H.G. Knuttgen, J.A. Vogel, and J. Poortmans), pp. 441–6. Human Kinetics Inc, Champaign, Illinois.

Chance, B., Sies, H., and Boveris, A. (1979). Hydroperoxide metabolism in mammalian organs. *Physiological Reviews*, **59**, 527–605.

Cotterill, L.A., Gower, J.D., Fuller, B.J., and Green, C.J. (1990). Free fatty acid accumulation following cold ischaemia in rabbit kidneys and the involvement of a calcium-dependent phospholipase A_2. *Cryoletters*, **11**, 3–12.

Crane, F.L., Sun, I.L., Clark, M.G., Grebing, C., and Low, H. (1985). Transplasma-membrane redox systems in growth and development. *Biochimica et Biophysica Acta*, **811**, 233–64.

Cullen, M.J. and Fulthorpe, J.J. (1982). Phagocytosis of the A band following Z line and I band loss. Its significance in skeletal muscle breakdown. *Journal of Pathology*, **138**, 129–43.

Curello, S., Ceconi, A., Cargnoni, A., Connachian, A., Ferrani, R., and Albertini, A. (1987). Improved procedure for determining glutathione in plasma as an index of myocardial oxidative stress. *Clinical Chemistry*, **33**, 1448–9.

Currie, R.W., Karmazyn, M., Kloc, M., and Mailer, K. (1988). Heat shock response is associated with enhanced postischaemic ventricular recovery. *Circulation Research*, **63**, 543–9.

Das, D.K., Engelman, R.M., Ronson, J.A., Breyer, R.H., Otani, H., and Lemeshow, S. (1985). Role of membrane phospholipids in myocardial injury induced by ischaemia and reperfusion. *American Journal of Physiology*, **251**, H71–9.

Davies, K.J.A., Quintanilha, A.T., Brooks, G.A., and Packer, L. (1982). Free radicals and tissue damage produced by exercise. *Biochimica et Biophysica Research Communications*, **107**, 1198–205.

De Leiris, J. and Feuvray, D. (1979). Morphological correlates of myocardial enzyme leakage. In *Enzymes in cardiology* (ed. D.J. Hearse and J. De Leiris), pp. 445–60. John Wiley and Sons, Chichester.

De Martino, G.N. and Croall, D.E. (1987). Calcium-dependent proteases: a prevalent proteolytic system of uncertain function. *News in Physiological Science*, **2**, 82–5.

Donnely, T.J., Sievers, R.E., Vissern, F.L.J., Welch, W.J., and Wolfe, C.L. (1992). Heat shock protein induction in rat hearts. A role for improved myocardial salvage after ischaemia and reperfusion. *Circulation*, **85**, 769–78.

Duncan, C.J. (1978). Role of intracellular calcium in promoting muscle damage: a strategy for controlling the dystrophic condition. *Experientia*, **34**, 1531–5.

Duncan, C.J. and Jackson, M.J. (1987). Different mechanisms mediate structural changes and intracellular enzyme efflux following damage to skeletal muscle. *Journal of Cell Science*, **87**, 183–8.

Duncan, C.J., Smith, J.L., and Greenway, H.C. (1979). Failure to protect frog skeletal muscle from ionophore-induced damage by the use of the protease inhibitor leupeptin. *Comparative Biochemistry and Physiology*, **63C**, 205–7.

Duncan, C.J., Greenaway, H.C., and Smith, J.L. (1980). 2,4 Dinitrophenol, lysosomal breakdown and rapid myofilament degradation in vertebrate skeletal muscle. *Naunyn-Schmiedebergs Archiv fur Experimentelle pathologie und pharmakologie*, **315**, 77–82.

Duthie, G.G. and Arthur, J.R. (1993). Free radicals and calcium homeostasis: relevance to malignant hyperthermia. *Free Radical Biology and Medicine*, **14**, 435–42.

Engel, W. King (1978). Integrative histochemical approach to the defect of Duchenne muscular dystrophy. In *Pathogenesis of human muscular dystrophies* (ed. L.P. Rowland), pp. 277–309. Excerpta Medica, Amsterdam.

Fantone, J.C. and Ward, P.A. (1985). Polymorphonuclear leukocyte-mediated cell and tissue injury: oxygen metabolites and their relation to human disease. *Human Pathology*, **16**, 973–8.

Font, B., Vial, C., and Goldschmidt, D. (1977). Metabolite levels and enzyme release in the ischaemic myocardium. *Journal of Molecular Medicine*, **2**, 291–7.

Garramone, R.R., Winters, R.M., Das, D.K., and Deckers, P.J. (1994). Reduction of skeletal muscle injury through stress conditioning using the heat shock response. *Plastic and Reconstructive Surgery*, **93**, 1242–7.

Gee, D.L. and Tappel, A.L. (1981). The effect of exhaustive exercise on expired pentane as a measure of in vivo lipid peroxidation in the rat. *Life Sciences*, **28**, 2425–9.

Gerner, E.W. and Schneider, M.J. (1975). Induced thermal resistance in Hela cells. *Nature*, **256**, 500–2.

Gillis, J.M. (1985). Relaxation of vertebrate skeletal muscle: a synthesis of the biochemical and physiological approaches. *Biochimica et Biophysica Acta*, **811**, 97–145.

Gohil, K., Viguie, C.A., Stanley, W.C., Packer, L., and Brooks, G.A. (1988). Blood glutathione oxidation during human exercise. *Journal of Applied Physiology*, **64**, 115–19.

Halliwell, B. and Gutteridge, J.M.C. (1984). Lipid peroxidation, oxygen radicals, cell damage and antioxidant therapy. *Lancet*, **ii**, 1396–7.

Halliwell, B. and Gutteridge, J.M.C. (ed.) (1985). *Free radicals in biology and medicine*. Clarendon Press, Oxford.

Harris, W.E. and Stahl, W.L. (1980). Organisation of thiol groups of electric eel organ Na/K ion stimulated adenosine triphosphatase, studied with bifunctional reagents. *Biochemical Journal*, **185**, 787–90.

Hearse, D.J. (1979). Cellular damage during myocardial ischaemia: metabolic changes leading to enzyme leakage. In *Enzymes in cardiology* (ed. D.J. Hearse and J. De Leiris), pp. 1–20. John Wiley and Sons, Chichester.

Hockenbery, D.M., Oltvai, Z.N., Yin, X.M., Milliman, C.L., and Korsmeyer, S.J. (1993). Bcl-2 functions in an antioxidant pathway to prevent apoptosis. *Cell*, **75**, 241–51.

Jackson, M.J. and Edwards, R.H.T. (1988). Free radicals, muscle damage and muscular dystrophy. In *Reactive oxygen species in chemistry, biology and medicine* (ed. A. Quintanilha), pp. 197–210. Plenum, New York.

Jackson, M.J., Jones, D.A., and Edwards, R.H.T. (1984a). Experimental skeletal muscle damage: The nature of the calcium-activated degenerative processes. *European Journal of Clinical Investigation*, **14**, 369–74.

Jackson, M.J., Jones, D.A., and Harris, E.J. (1984b). Inhibition of lipid peroxidation in skeletal muscle homogenates by phospholipase A_2 inhibitors. *Bioscience Reports*, **4**, 581–7.

Jackson, M.J., Edwards, R.H.T., and Symons, M.C.R. (1985). Electron spin resonance studies of intact mammalian skeletal muscle. *Biochimica et Biophysica Acta*, **847**, 185–90.

Jackson, M.J., McArdle, A., and Edwards, R.H.T. (1991a). Free radicals, calcium and damage in dystrophic and normal skeletal muscle. In *Calcium free radicals and tissue damage* (ed. C.J. Duncan), pp. 139–48. Cambridge University Press.

Jackson, M.J., Kaiser, K., Brooke, M.H., and Edwards, R.H.T. (1991b). Glutathione depletion during experimental damage to skeletal muscle and its relevance to Duchenne muscular dystrophy. *Clinical Science,* **80**, 559–64.

Johnson, K., Sutcliffe, L., Edwards, R.H.T., and Jackson, M.J. (1988). Calcium ionophore enhances the electron spin resonance signal from isolated skeletal muscle. *Biochimica et Biophysica Acta,* **964**, 285–8.

Jones, D.A., Jackson, M.J., and Edwards, R.H.T. (1983). The release of intracellular enzymes from an isolated mammalian skeletal muscle preparation. *Clinical Science,* **65**, 193–201.

Jones, D.A., Jackson, M.J., McPhail, G., and Edwards, R.H.T. (1984). Experimental muscle damage: the importance of external calcium. *Clinical Science,* **66**, 317–22.

Karthuis, R.J., Granger, D.N., Townsley, M.I., and Taylor, A.R. (1985). The role of oxygen-derived free radicals in ischaemia-induced increases in canine skeletal muscle vascular permeability. *Circulation Research,* **57**, 599–609.

Lam, M., Dubyak, G., Chen, L., Nunez, G., Miesfield, R.L., and Distelhorst, C.W. (1994). Evidence that Bcl-2 represses apoptosis by regulating endoplasmic reticulum associated Ca^{2+} fluxes. *Proceedings of the National Academy of Sciences USA,* **91**, 6569–73.

Li, G.C. and Werb, Z. (1982). Correlation between synthesis of heat shock proteins and development of thermotolerance in chinese hamster fibroblasts. *Proceedings of the National Academy of Sciences USA,* **79**, 3219–22.

Marber, M.S. (1994). Stress proteins in myocardial protection. *Clinical Science,* **86**, 375–81.

McArdle, A., Edwards, R.H.T., and Jackson, M.J. (1991). Effects of contractile activity on muscle damage in the dystrophin-deficient *mdx* mouse. *Clinical Science,* **80**, 367–71.

McArdle, A., Edwards, R.H.T., and Jackson, M.J. (1992). Accumulation of calcium by normal and dystrophin-deficient muscle during contractile activity 'in vitro'. *Clinical Science,* **82**, 455–9.

McArdle, A., Edwards, R.H.T., and Jackson, M.J. (1993). Calcium homeostasis during contractile activity of vitamin E deficient skeletal muscle. *Proceedings of the Nutrition Society,* **52**, 83A.

McCord, J.M. (1985). Oxygen-derived free radicals in post-ischaemic tissue injury. *New England Journal of Medicine,* **312**, 159–63.

Mellgren, R.L. (1987). Calcium dependent proteases, an enzyme system active at cellular membranes. *Federation of American Societies for Experimental Biology (FASEB) Journal,* **1**, 110–15.

Michell, R.H. and Coleman, R. (1979). Structure and permeability of normal and damaged membranes. In *Enzymes in cardiology* (ed. D.J. Hearse and J. De Leiris), pp. 59–80. John Wiley and Sons, Chichester.

Murphy, M.E. and Kehrer, J.P. (1986). Free-radicals: A potential pathogenic mechanism in inherited muscular dystrophy. *Life Sciences,* **39**, 2271–8.

Nicotera, P.L., Moore, M., Mirabelli, F., Bellomo, G., and Orrenius, S. (1985). Inhibition of hepatocyte plasma membrane Ca^{2+} ATPase activity by menadione metabolism and its restoration by thiols. *Federation of European Biochemical Societies (FEBS) Letters,* **181**, 149–53.

Packer, L. (1984). Vitamin E, physical exercise and tissue damage in animals. *Medical Biology,* **62**, 105–9.

Pennington, R.F.T. (1988). Biochemical aspects of muscle disease. In *Disorders of voluntary muscle* (ed. J. Walton), pp. 455–86. Churchill Livingstone, Edinburgh.

Phoenix, J., Edwards, R.H.T., and Jackson, M.J. (1990). Effects of calcium ionophore on vitamin E deficient muscle. *British Journal of Nutrition*, **64**, 245–56.

Polla, B.S., Mini, N., and Kahtengwa, S. (1991). Heat shock and oxidative injury in human cells. In *Heat shock* (ed. B. Maresca and D. Lindquist), pp. 279–90. Springer-Verlag, Berlin.

Publicover, S.J., Duncan, C.J., and Smith, J.L. (1978). The use of A23187 to demonstrate the role of intracellular calcium in causing ultrastructural damage in mammalian muscle. *Journal of Neuropathology and Experimental Neurology*, **37**, 554–7.

Riabowol, K.T., Mizzen, L.A., and Welch, W.J. (1988). Heat shock is lethal to fibroblasts microinjected with antibodies against HSP70. *Science*, **242**, 433–6.

Robertson, J.D., Maughan, R.J., Duthie, G.G., and Morrice, P.C. (1991). Increased blood antioxidant systems of runners in response to training load. *Clinical Science*, **80**, 611–18.

Rodemann, H.P. and Goldberg, A.L. (1982). Arachidonic acid, prostaglandin E_2 and F_{2a} influence rates of protein turnover in skeletal and cardiac muscle. *Journal of Biological Chemistry*, **257**, 1632–8.

Rodemann, H.P., Waxman, L., and Goldberg, A.L. (1982). The stimulation of protein degradation in muscle by Ca^{2+} is mediated by prostaglandin E_2 and does not require the calcium-activated protease. *Journal of Biological Chemistry*, **257**, 8716–23.

Schanne, F.X., Kane, A.B., Young, A.B., and Forber, J.C. (1979). Calcium dependence of toxic cell death: a final common pathway. *Science*, **206**, 700–1.

Schatzmann, H.J. (1975). Active calcium transport and Ca^{2+}-activated ATPase in human red cells. *Current Topics in Membranes and Transport*, **6**, 125–68.

Scherer, N.M. and Deamer, E.W. (1986). Oxidative stress impairs the function of sarcoplasmic reticulum by oxidation of sulphydryl groups in the Ca^{2+} ATPase. *Archives of Biochemistry and Biophysics*, **246**, 589–601.

Schwartz, L.M. and Osborne, B.A. (1993). Programmed cell death, apoptosis and killer genes. *Immunology Today*, **14**, 582–90.

Schwartz L.M., Kosz. L., and Kay, B.K. (1990). Gene activation is required for developmentally programmed cell death. *Proceedings of the National Academy of Sciences USA*, **87**, 6594–8.

Schwartz, L.M., Jones, M.E.E., Kosz, L., and Kaah, K. (1993). Selective repression of actin and myosin heavy chain expression during the programmed death of insect skeletal muscle. *Developmental Biology*, **158**, 448–55.

Soybell, D., Morgan, J., and Cohen, L. (1978). Calcium augmentation of enzyme leakage site of action. *Research Communications in Chemical Pathology and Pharmacology*, **70**, 17–329.

Thomas, J.A., Chai, Y.C., and Jung, C.H. (1994). Protein S-thiolation and dethiolation. In *Methods in enzymology, oxygen radicals in biological systems*, Vol. 233. (ed. L. Packer), pp. 385–97. Academic Press, London.

Tidball, J.G., Albrecht, D.E., Lokensgard, B.E., and Spencer, M.J. (1995). Apoptosis precedes necrosis of dystrophin-deficient muscle. *Journal of Cell Science*, **108**, 2197–204.

West-Jordan, J.A., Martin, P.A., Abraham, R.J., Edwards, R.H.T., and Jackson,

M.J. (1990). Energy dependence of cytosolic enzyme efflux from rat skeletal muscle. *Clinica Chimica Acta,* **189**, 163–72.

West-Jordan, J.A., Martin, P.A., Abraham, R.J., Edwards, R.H.T., and Jackson, M.J. (1991). Energy metabolism during damaging contractile activity in isolated skeletal muscle: A ^{31}P-NMR study. *Clinica Chimica Acta,* **203**, 119–34.

Wrogemann, K. and Pena, S.J.G. (1976). Mitochondrial overload—a general mechanism for cell necrosis in muscle diseases. *Lancet,* **ii**, 672–4.

Yellon, D.M., Pasini, E., Cargoni, A., Marber, M.S., Latchman, D.S., and Ferrari, R. (1992). The protective role of heat stress in the ischaemic and reperfused myocardium. *Journal of Molecular and Cellular Cardiology,* **24**, 895–907.

Zerba, E., Komorowski, T.E., and Faulkner, J.A. (1990). Free radical injury to skeletal muscles of young, adult and old mice. *American Journal of Physiology,* **258**, C429–35.

Molecular markers of muscle plasticity, damage, regeneration, and repair

Corrado Rizzi, Francesco Mazzoleni, Marco Sandri, Katia Rossini, Adonella Bruson, Marco Rossi, Claudia Catani, Marcello Cantini, and Ugo Carraro

7.1 Introduction

The motor units of skeletal muscles differ in their ability to sustain a given power output without stress and fatigue. They may therefore be damaged by different levels of exercise, as well as by events involving ischaemia and reperfusion, or administration of myotoxic agents. Quantitation of such damage is often performed morphometrically, but such an approach is vulnerable to the artefacts that can appear in cryostat sections taken from seriously disrupted tissue. Paraffin and acrylic embedding for light and electron microscopy are laborious procedures, and require substantial biopsies whose quantitation can be time-consuming (see Chapter 10).

In this chapter we will describe scaled-down procedures, based on protein and nucleic acid chemistry, that are sensitive enough to allow several markers of plasticity, damage, regeneration, and repair to be determined on a muscle sample of just a few milligrams (Carraro *et al.* 1985). A suitable source of material is a small biopsy specimen removed from the muscle for fibre typing and histopathology; a few additional serial cryostat sections cut from the block constitute a sample that is adequate for biochemical analysis. Since such biopsies can be taken sequentially, it becomes feasible to construct a time course for the response of skeletal muscle to different demands in either experimental or clinical settings.

We have validated micromethods for protein and collagen content, myosin:actin ratio, and myosin heavy chain (MHC) composition. Total protein and collagen may be used as indices of muscle plasticity, regeneration, and repair; the myosin:actin ratio is a reliable index of the onset of muscle atrophy; MHC composition is an indicator of regenerative events and of adaptation to changing demands.

Where indicated, these protocols may be combined with other analyses, such as the identification of proteins with antibodies or *in situ* hybridization with specific nucleic acid markers. A variety of highly sensitive detection methods may be added in this way: for example, chemiluminescence and polymerase chain reaction techniques are capable of detection sensitivities that extend almost to the level of a single molecule (Bronstein *et al.* 1992). The procedures can be adapted to the characterization of any tissue by selecting the appropriate molecular markers. In the case of muscle, information obtained from needle biopsy material can be supplemented by determinations of myoglobin and creatine kinase (CK) in the serum.

We have developed a number of standard protocols for the molecular characterization of muscle biopsies in different scientific or clinical contexts. Examples are:

- dystrophin in Duchenne muscular dystrophy (Hoffman *et al.* 1987);
- inflammatory cell markers in chronic myosites;
- mutant MHC in familial hypertrophic cardiomyopathy (Dausse *et al.* 1993);
- mitochondrial or glycolytic enzymes in mitochondrial myopathies or glycogenoses;
- fragmented DNA from apoptotic myonuclei (Podhorska-Okolow *et al.* 1995; Sandri *et al.* 1995);
- protein or isotopic markers in exercise-induced muscle damage (see Chapters 1, 5, 8, and 9);
- isomyosin transitions in skeletal muscle-powered circulatory assist devices (Carraro and Arpesella 1991; Stephenson 1993).

To obtain the best correspondence between conventional histopathological methods and the molecular characterization based on these analyses, the latter needs to be performed on serial sections cut from the same frozen muscle samples. This poses the problem of distortion and shrinkage of the sections as they adhere to and dry on glass slides, which would of course affect calculations of the absolute content of the various components. A simple solution to this problem is described, and this allows us to reproduce the results that are obtained when standard biochemical approaches are applied to larger samples of the same tissue.

7.2   Morphometry

7.2.1   Determination of the area and volume of cryostat sections

When the entire muscle is available for analysis, it is held at resting length, frozen quickly in liquid nitrogen, and stored at $-80°C$ until needed. The sample for analysis is taken from midbelly. Serial sections are cut from these samples, or from samples obtained by biopsy, in a cryostat. For histology or immunohistochemistry, sections of 10 μm thickness are collected on glass slides; for biochemical analysis, sections of 20 μm thickness are transferred to conical glass test tubes. The slides and test tubes are usually stored at −80°C to await analysis. After the sections have been taken, the face of the muscle block still mounted in the cryostat is photographed alongside a standard microgrid. At a later stage the slides can be projected, and from a knowledge of the grid size the cross-sectional area can be determined. Since the cryostat sections are of known thickness, their volume is readily calculated.

For whole muscles, the area of the sectioned face of the cryostat-mounted block (Fig. 7.1) corresponds to the actual cross-section of the muscle belly *in vivo*, and provides a good measure of muscle mass (Fig. 7.2). This is not true

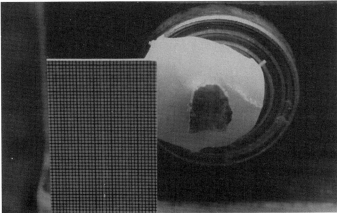

Fig. 7.1 Technique for measuring the area of cryostat sections. (A) Photographic set-up. (B) Sectioned face of a cryostat muscle block and the reference grid. The area is calculated by projecting the slide and superimposing the block face on the mesh, or by scanning the picture and performing computer-based measurements.

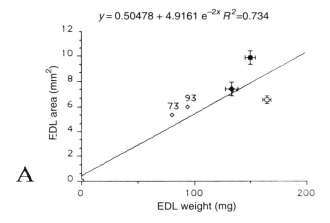

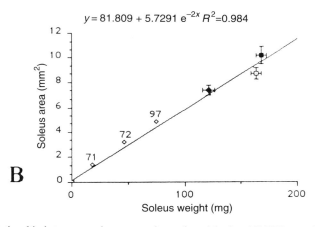

Fig. 7.2 Relationship between section area and muscle weight for: (A) EDL muscle; (B) soleus muscle. 73 = cross-transplanted soleus, experiment number 73; 93 = cross-transplanted soleus, experiment number 93; 72 = cross-transplanted EDL, experiment number 72; 71 = cross-transplanted EDL, experiment number 71; 97 = cross-transplanted EDL, experiment number 97. (Filled circles) Mean ± SEM of the section areas and muscle weights for the 250 g rats; (open squares) mean ± SEM of the section areas and muscle weights for the 350 g rats; (filled squares) mean ± SEM of the section areas and contralateral control muscle weights for the experimental rats.

of the area of cryostat sections that are cut from the block and mounted on glass slides, because of shrinkage during adhesion and drying. For example, the mean section area of cryostat blocks obtained from the normal extensor digitorum longus (EDL) muscles of large rats was found to be 6.5 ± 0.3 mm^2, whereas the area of the sections mounted on slides had decreased to 4.76 ± 0.16 mm^2 (mean ± SEM, n = 12 muscles, $p < 0.0001$).

Provided that the overall shape of the muscles is the same, the ratio of section area to muscle weight may be used in comparisons as an index of muscle hypotrophy or hypertrophy. Where, as in the case of muscle biopsies, it is not possible to freeze the whole muscle at resting length, the parameter is of less value as an index of muscle girth. Section area is still worth determining, however, as it allows the concentration of a given muscle component to be calculated.

The use of these techniques, and techniques to be described below, will be illustrated by reference to the following experimental material. Table 7.1 summarizes data on normal rat muscles, taken from two groups of animals in which the average body weights were 250 g and 350 g. Table 7.2 presents data from the control and experimental sides of rats in which EDL and soleus muscles were removed, treated with bupivacaine (a myotoxic agent), and transplanted as free grafts, each muscle into the bed of the other. Values from normal and contralateral control muscles were unimodally distributed, and data are therefore presented as mean ± SEM. Results from the transplanted muscles are presented individually because of the large variation they showed in the extent of fibre regeneration and reinnervation.

7.2.2 Fibre typing and interstitial tissue content

Serial sections (10 μm) are stained by the conventional techniques for haematoxylin and eosin (H & E) and haematoxylin-Van Gieson, and for myofibrillar ATPase according to the method of Brooke and Kaiser (1970). The ATPase sections are preincubated at alkaline or acid pH and used to measure the size of type 1 and type 2 fibres. Morphometric analyses are performed on glass slides or photographs with an image analyser (Kontron IBAS 2000, Zeiss). The section profile is imaged by a television scanner and grey level detector, which facilitates identification, selection, and measurement of single features (Mussini *et al.* 1987). Fibre counts are based on all the fibres identified in H & E-stained sections. The percentage content of type 1 fibres is determined in randomly selected areas of ATPase-stained sections. In experimental muscles it may not be easy to classify type 2 fibres into the traditional subgroups, because some fibres are stained with intermediate intensity after acid and alkaline preincubation (see, for example, Fig. 7.3). These intermediate fibres are often encountered in muscles undergoing transformation, for example during long-term electrical stimulation of either innervated muscles (Salmons and Henriksson 1981; Carraro *et al.* 1988; Ausoni *et al.* 1990; Carraro and Arpesella 1991; Mayne *et al.* 1993) or denervated muscles (Carraro *et al.* 1986; Ausoni *et al.* 1990). They are hybrid fibre types that contain more than one isoform of myosin (Pette and Staron 1990; Pette and Vrbová 1992).

Table 7.1 Morphological markers of growth, plasticity, damage, regeneration, and repair of skeletal muscle in the adult rat

	Normal soleus		Free cross-transplantation of buvicaine-treated muscles*							Normal EDL	
			Contralateral	x-EDL (exp. no.)			x-sol (exp. no.)		Contralateral		
	250 g rat	350 g rat	soleus	71	72	97	93	73	EDL	350 g rat	250 g rat
Body weight (g)	267 ± 5	352 ± 4	384 ± 16	422	373	343	367	440	384 ± 16	352 ± 4	267 ± 5
Muscle weight (g)	121 ± 4	164 ± 4	168 ± 3	47	18	75	912	75	150 ± 4	165 ± 4	133 ± 3
Muscle/body weight ratio	0.45 ± 0.01	0.46 ± 0.01	0.44 ± 0.16	0.11	0.05	0.22	0.25	0.17	0.40 ± 0.00	0.47 ± 0.01	0.50 ± 0.01
Section area (mm^2)	7.4 ± 0.2	8.7 ± 0.4	10.2 ± 0.7	3.2	1.4	4.9	6.0	5.5	10.0 ± 0.6	6.5 ± 0.3	7.4 ± 0.5
Collagen area (%)	—	—	0.8 ± 0.3	10.3	4.1	10.7	9.5	11.8	0.05 ± 0.1	—	—
Fibre number	2201 ± 104	3214 ± 142	2038 ± 94	975	1075	3105	896	714	3039 ± 361	3257 ± 129	3601 ± 297
Type 1 fibres (%)	89.2 ± 1.2	71.26 ± 1.72	91.2 ± 2.6	3.4	30.3	1.4	2.0	0.3	3.2 ± 0.8	1.03 ± 0.2	3.3 ± 0.4

* exp. no., Experiment number.

Table 7.2 Molecular markers of growth, plasticity, damage, regeneration, and repair of skeletal muscle in the adult rat

	Normal soleus		Free cross-transplantation of buvicaine-treated muscles*							Normal EDL	
			Contralateral	x-EDL (exp. no.)			x-sol (exp. no.)		Contralateral		
	250 g rat	350 g rat	soleus	71	72	97	93	73	EDL	350 g rat	250 g rat
Protein (µg) per section	26.3±1.6	34.8±2.0	41.1±2.0	9.0	5.2	8.4	20.0	7.0	42.9±1.7	33.5±1.9	28.8±1.8
Protein (%)	17.3±0.8	20.2±1.06	20.5±1.07	14.1	18.3	8.6	16.7	6.4	21.9±1.9	26.2±1.8	19.9±1.4
Collagen (µg) per section	1.6±0.1	1.0±0.1	1.0±0.1	0.4	0.2	1.2	3.7	3.8	0.8±0.1	0.8±0.1	2.2±0.3
Collagen (%)	1.0±0.1	0.6±0.1	0.5±0.1	0.6	0.7	1.3	3.1	3.5	0.4±0.1	0.6±0.1	1.5±0.1
Collagen/protein ratio ×100	5.9±0.7	3.1±0.3	2.4±0.3	3.8	3.6	14.7	18.6	54.5	1.8±0.1	2.4±0.3	7.7±1.4
Myosin/actin	2.3±0.2	2.2±0.04	2.3±0.1	1.7	1.2	1.5	1.1	1.6	1.8±0.5	2.1±0.1	1.9±0.1
MHC2A + 2X (%)	8.4±1.2	26.1±1.4	11.3±2.8	79.0	58.0	39.0	75.0	73.5	47.3±2.5	36.3±1.4	35.5±3.2
MHC2B (%)	0	0	4.7±3.0	16.0	0	58.0	19.0	24.0	46.7±1.8	62.0±1.4	64.0±3.3
MHC1 (%)	91.6±1.2	73.9±1.4	84.1±5.4	5.0	42.0	3.0	6.0	2.5	4.8±1.0	0.65±0.4	0.5±0.3

* exp. no., Experiment number.

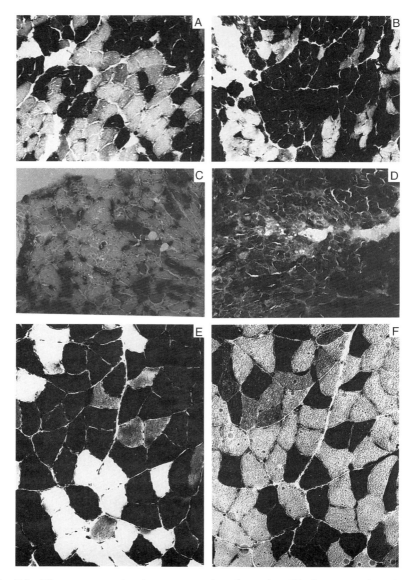

Fig. 7.3   Fibre type grouping in cross-transplanted muscles. (A) Cross-transplanted EDL, experiment number 72; (B) cross-transplanted EDL, experiment number 72; (C) cross-transplanted soleus, experiment number 93; (D) cross-transplanted EDL, experiment number 93; (E) contralateral control soleus, experiment number 71; (F) contralateral control soleus, experiment number 71. (A), (C), (E) ATPase pH 4.35; (B), (D), (F) ATPase pH 10.4. Magnification, 117 × .

In the normal and contralateral control soleus and EDL muscles there were about 3000 fibres (Table 7.1). The muscles contained the expected complement of type 1 fibres: about 3 per cent in EDL and 90 per cent in soleus. Type-grouping was evident in the transplanted muscles, indicating that most of the fibres had been reinnervated. The total area of connective tissue was less than 1 per cent in the normal and contralateral control EDL and soleus muscles; in the cross-transplanted muscles, it had increased as much as 10–fold (Table 7.1).

7.3   Molecular markers

7.3.1   Total protein

Two cryostat sections are collected in cylindrical glass test tubes and are dissolved in 0.5 per cent deoxycholic acid by sonication at room temperature (Biosonik III, Bronwill Scientific). The protein content is determined by the method of Lowry *et al.* (1951).

Skeletal muscle normally contains about 20 per cent of total protein by wet weight, a result of the high content of contractile proteins in the fibres (Szent-Gyorgyi 1945). The value obtained by cryostat section analysis of normal and contralateral muscles in the experimental material was 21.93 ± 0.76 per cent (mean $\pm$ SEM), with minor variation between the EDL and soleus muscle groups (Table 7.2).

In determining total protein from sections a problem arises if the medium used to embed the frozen tissue specimen is included with the sections. This material interferes with the assay, whether by Lowry or Bradford (1976) methods, and such contamination must therefore be carefully avoided. An additional problem is encountered in experimental or pathological conditions that involve an increase in the collagen content of the muscle, since collagen does not register in the Lowry method and is underestimated with the Bradford method; this is because it contains very low amounts of the aromatic amino acids that are the basis for the colorimetric reaction (Wilson 1979).

7.3.2   Collagen

Two serial sections are collected in a pluggable Pyrex test tube and analysed for total collagen by a modification of the method described in Woessner (1961) and Huszar *et al.* (1980), as follows. The sections are hydrolysed for 20 h in 1 ml of 6N HCl at 110°C. After hydrolysis the samples are dried either in a rotary evaporator or in a multi-sample bench-top centrifuge connected to a water-driven vacuum pump (Electrofor, Rovigo, Italy).

Dried samples are dissolved in 1 ml of distilled water. Aliquots of 0.4 ml are added to 0.5 ml of chloramine-T solution; this is prepared by dissolving 1.41 g sodium N-chloro-p-toluene sulfonamide in 10 ml n-propanol, 10 ml water, and 80 ml of a pH 6 buffer. The pH 6 buffer is prepared by dissolving 25 g citric acid monohydrate, 60 g sodium acetate trihydrate, 17 g NaOH, and 6 ml glacial acetic acid in distilled water to a final volume of 500 ml. The sample is mixed vigorously in the chloramine-T solution and incubated at room temperature (20–25°C). After 20 min, 0.77 ml of a freshly prepared aldehyde/perchloric acid solution is added; this is made up by solubilizing 15 g p-dimethylaminobenzaldehyde in 62 ml n-propanol to which 26 ml 60 per cent perchloric acid are slowly added, followed by distilled water to a total volume of 100 ml. The samples are vortexed and incubated at 65°C for 20 min. Full colour development occurs in 20 min. The samples are read at 550 nm, usually straight away, although this is not essential as the colour is stable for several hours. Hydroxyproline content is determined by interpolation from a hydroxyproline standard curve made on each occasion. Under the conditions described, the curve is linear from 0.5 to 20 µg of hydroxyproline. Since the mean content of hydroxyproline in collagen is 13.4 per cent, collagen is easily calculated, due account being taken of the lower molecular weight of the amino acid in polypeptides, in which it is linked by peptide bonds.

The concentration of collagen usually increases in muscles undergoing tissue repair when necrosis is due to physical trauma, and when ischaemic and/or toxic necrosis is complicated by infection. Collagen content may also change with muscular activity: for example, morphological studies have demonstrated a considerable increase in the proportion of connective tissue in muscle immobilized in a lengthened position, and a decrease during remobilization, indicating that the biosynthesis of collagen in muscle adapts to stretch or activity (Savolainen *et al.* 1987). The collagen:protein ratio varies even among normal muscles, because variation in total protein content is amplified by the failure of the Lowry method for protein determination to include collagen. Nevertheless, the collagen:protein ratio, like the total protein content, provides a useful index of functional integrity.

In our experiments, normal and contralateral muscles had collagen contents close to the expected value of 0.7 per cent for rat muscles (Savolainen *et al.* 1987). The collagen concentration had increased in cross-transplanted soleus muscles, but the figures for cross-transplanted EDL muscles were not significantly different from controls. Notwithstanding this last observation, the percentage area occupied by interstitial tissue had increased more than 10–fold in the cross-transplanted muscles (Table 7.1). Because the cross-transplanted muscles had a low content of contractile proteins, the collagen: protein ratio increased substantially— up to 10–fold in cross-transplanted soleus muscles (Table 7.2)—indicative of a very limited functional recovery of the transplanted muscles. This is an important conclusion to derive from such small biopsy samples. Further evidence of atrophy in the cross-transplanted muscles came from their low myosin:actin ratio (see below).

7.3.3 Contractile proteins

Two serial sections are solubilized at 100 mg/ml in 2.3 per cent sodium dodecyl sulphate (SDS), 10 per cent (by volume) glycerol, 0.5 per cent 2–mercaptoethanol, 62.5 mM Tris HCl, pH 6.8, based on their protein content determined by the Lowry method. Test tubes are closed with aluminium foil, and incubated for 3 min in boiling water to denature the proteins.

7.3.3.1 Myosin: actin ratio

Analytical SDS polyacrylamide gel electrophoresis (SDS-PAGE) is performed essentially according to Laemmli (1970) in a discontinuous gel gradient system within a $0.75 \times 130 \times 130$ mm polyacrylamide gel slab: this comprises a 10–mm stacking gel and a separating gel consisting of 90 mm of 12.5 per cent polyacrylamide and 30 mm of 7 per cent polyacrylamide. A major modification is that 37.5 per cent (by volume) glycerol is included in both separating and stacking gels (Carraro and Catani 1983; Danieli-Betto *et al.* 1986). The electrophoresis buffer consists of 25 mM Tris, 192 mM glycine pH 8.3, and 0.1 per cent SDS. Overnight separation is achieved in constant-current mode at a current of about 5 mA per slab, the actual current being set to a level that gives an initial voltage of 35 V. Gel electrophoresis is started late in the afternoon and the run is terminated after about 18 h when the dye front reaches the end of the slab. The voltage usually rises to 140–150 V by the end of the run. Normally 5 μg of total protein are loaded per well. MHC and actin bands are easily detected after shaking for 1h with 200 ml of 0.1 per cent Coomassie Brilliant Blue in 5 per cent acetic acid, 40 per cent methanol and destaining in 6 changes (every 15 min) of 200 ml 40 per cent methanol, 7 per cent acetic acid. To avoid variability in the stain-destain procedure the MHC:actin ratio is checked against a standard muscle homogenate that is run with each gel. MHC and actin content are determined by gel densitometry, as described later.

Myosin and actin have different rates of metabolic turnover, and the myosin:actin ratio changes during muscle atrophy and regrowth, for example, after denervation-reinnervation (Jakubiec-Puka *et al.* 1981, 1982) or tenotomy-repair (Jakubiec-Puka *et al.* 1992). Figure 7.4 shows the patterns of contractile proteins on the discontinuous 7/12.5 per cent polyacrylamide gradient. The discontinuous gradient gel is needed to resolve MHC and actin from proteins that migrate very closely under normal gel electrophoresis conditions, particularly in the region of the actin band in the case of soleus muscle. An electrophoretic approach similar to that used to calculate the myosin:actin ratio may be used to determine the myoglobin content of muscle, which provides an index of the oxidative metabolic capacity of the tissue.

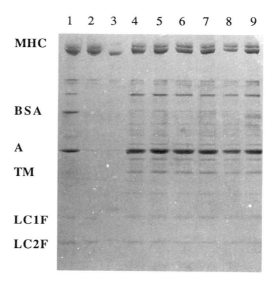

Fig. 7.4 SDS 7/12.5 per cent PAGE of contractile proteins for the determination of myosin:actin ratio. Coomassie Brilliant Blue stain. 1, 3 µg of total protein from homogenate of gastrocnemius supplemented with 0.5 µg of purified albumin; 2, 2 µg of myosin purified from EDL; 3, 1 µg of myosin purified from soleus; 4–9, 5 µg of total protein from cryostat sections of EDL from a rat of body weight 250 g. Myosin:actin ratios were determined by densitometry as described in the text.

In the illustrative experiments, the myosin:actin ratio obtained by densitometry of the gel slabs had the expected mean value of 2.2 in normal and contralateral control muscles. In agreement with histological evidence of hypotrophy (Fig. 7.3), the ratio decreased to values typical of atrophying muscle in the cross-transplanted muscle grafts. These muscle grafts contained many small angular fibres. By analogy with changes in the size and myosin composition of regenerated muscle fibres in innervated or permanently denervated limbs (Carraro *et al.* 1983; Mussini *et al.* 1987), these could be interpreted as fibres that initially underwent regeneration but then atrophied because they were not reinnervated. Small fibres that could be type-grouped by ATPase staining were probably reinnervated fibres, their small size reflecting either an early stage of growth or ongoing disuse atrophy.

7.3.3.2 *Myosin heavy chain isoforms*

MHC are separated by analytical SDS-PAGE on 7 per cent polyacrylamide gel slabs measuring $0.75 \times 130 \times 130$ mm (Carraro and Catani 1983). The slabs are prepared according to Laemmli (1970), except that 37.5 per cent (by volume) glycerol is present in both separating and stacking gels (Danieli-Betto *et al.* 1986). The stacking gel is composed of 37.5 per cent glycerol, 4 per cent T acrylamide-Bis (36.5:1), 125 mM Tris HCl (pH 6.8), and 0.1 per cent SDS. The separating gel is composed of 37.5 per cent glycerol, 7 per cent T acrylamide-Bis (36.5:1), 375 mM Tris HCl (pH 8.8), and 0.1 per cent SDS. The running buffer consists of 50 mM Tris, 384 mM glycine pH 8.3, and 0.2 per cent SDS (without correcting the pH with HCl). Separation of MHC is achieved in the constant-current mode at 4 mA per slab, corresponding to a voltage of about 40 V. After 4–6 h the buffer is changed and electrophoresis is restarted with the same parameters. Usually the run is started in the morning; after 24 h the voltage rises to 130–160 V and the run is then stopped. Gels containing 0.2 µg of protein per band are stained with 0.1 per cent Coomassie Brilliant Blue in 5 per cent acetic acid, 40 per cent methanol and destained in 40 per cent methanol, 7 per cent acetic acid. Gels with less than 0.1 µg of protein per band are stained by the silver method (Merrill *et al.* 1981), modified according to Betto (1993).

For Western blotting, MHC are transferred in 25 mM Tris, 193 mM glycine, 20 per cent methanol for 5 h at 4°C at a constant voltage (80 V). Ponceau Red staining is used after blotting as a check on the amount and position of MHCs transferred. Immunostaining is then performed according to Cantini *et al.* (1993) with an antibody to MHC_{emb} (G6 clone), a generous gift from Professor Stefano Schiaffino.

Figure 7.5 shows the MHC patterns of our normal and experimental muscles. Quantitative densitometry of MHC separated on 7 per cent polyacrylamide (Table 7.2) revealed in normal soleus of 250 –g rats the expected values of 91.6 ± 1.2 per cent for MHC1 and 8.4 ± 1.2 per cent for MHC2A (mean ± SEM). The figures for normal EDL muscles of 35.5 ± 3.2 per cent for MHC 2A + 2X, 64.0 ± 3.2 per cent for MHC2B, and 0.5 ± 0.3 per cent for MHC1 were also in good agreement with published data (Carraro and Catani 1983; Bär and Pette 1988; Carraro *et al.* 1990; Jakubiec-Puka *et al.* 1990; La Framboise *et al.* 1990; Sugiura and Murakami 1990; Takahashi *et al.* 1991; Rizzi and Carraro 1991; Sandri *et al.* 1992; Talmadge and Roy 1993; Rossini *et al.* 1995; Table 7.3).

Earlier cross-innervation studies (Thomas and Ranatunga 1993) would have led us to expect a complete transformation of fibre type in the cross-transplanted muscles. The MHC patterns fell short of this predicted outcome. In the EDL muscle that had been transplanted to the soleus bed, a large to very large proportion of the MHC consisted of MHC2A or 2X. This muscle never attained the MHC1 content typical of the normal soleus, although the slow isoform was present in much larger amounts than in the normal EDL. A substantial MHC2B component was still seen in two out of the three experimental muscles. The soleus muscle transplanted to the EDL bed had lost almost all of its complement of MHC1 (3–6 per cent), and showed an increase in the content of MHC2A + 2X (74–75 per cent), particularly MHC2X. The MHC2B content (19–24 per cent) was much lower than in normal EDL muscles.

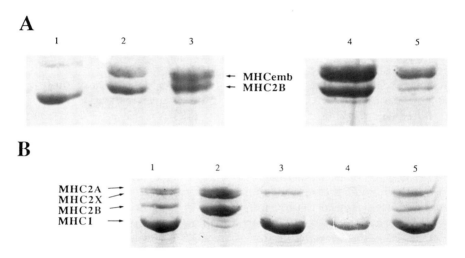

Fig. 7.5 High-resolution SDS-PAGE of myosin heavy chains on 7 per cent gels; silver stain. (A) 1, 1.25 µg of normal soleus; 2, 1.5 µg of normal EDL; 3, 2 µg of embryonic muscle; 4, 5 µg of cross-transplanted EDL, experiment number 97; 5, 1.25 µg of cross-transplanted soleus, experiment number 93. (B) 1, 2.5 µg of contralateral control soleus, experiment number 97; 2, 3 µg of contralateral control EDL, experiment number 92; 3, 2 µg of contralateral control soleus, experiment number 92; 4, 1 µg of contralateral control soleus, experiment number 72; 5, 2 µg of contralateral control soleus, experiment number 71. Myosin heavy chain isoforms MHC2A, MHC2X, MHC2B, and MHC1 are identified. Separation of MHC2A from MHC2X depends on their relative abundance.

Reference has already been made (§7.2.2) to the type-grouping found when the regenerated fibres in the cross-transplanted muscles were examined morphologically. The same observation could have been made by biochemical analysis of single fibres, but only at the expense of a good deal of tedious and time-consuming work (Pette and Staron 1990). This illustrates the value of the combined approach.

Silver staining of electrophoretograms prepared from transplanted muscles revealed trace amounts of the embryonic MHC isoform, MHC_{emb} (Rossini *et al.* 1995). Quantitation of this component could have been improved by chemiluminescent detection, for which commercial products are available (McCapra 1970; Voyta *et al.* 1988; Bronstein *et al.* 1989, 1992).

MHC$_{emb}$ can be used as a marker of skeletal muscle regeneration in adult muscle, if the timing is appropriate. The immunochemical analysis can be performed routinely on muscle biopsies, although the results can be compromised by cross-reactivity of the antibodies between MHC$_{emb}$ and adult fast MHC isoforms. In the next section we describe a peptide mapping approach to MHC$_{emb}$ identification, based on SDS-PAGE of hydroxylamine-cleaved contractile proteins, that may overcome problems of cross-reactivity.

7.3.3.3 *Hydroxylamine fragments of embryonic myosin heavy chains*

Hydroxylamine digestion of MHC is performed essentially according to Bornstein and Balian (1970), the method having been adapted to cryostat section handling by Sandri *et al.* (1993).

When MHC and their hydroxylamine digests are separated electrophoretically and immunoblotted with the anti-MHC$_{emb}$ antibody, a 55 kD peptide is revealed in both rat and human embryonic muscle. Hydroxylamine digestion of total proteins from cryostat sections produces SDS-PAGE patterns that are very complex, and these become virtually unreadable when they are overloaded to reveal minor components. Immunoblotting identifies the 55 kD peptide that is unique to MHC$_{emb}$ in 4–day, but not in 2–day, regenerating muscle (Fig. 7.6). This observation is consistent with the known kinetics of myotube development and embryonic myosin expression during bupivacaine-induced muscle regeneration (Hall-Craggs 1978; Carraro *et al.* 1983; Mussini *et al.* 1987; Cantini and Carraro 1993).

In our experiments the 55 kD peptide could not be detected in the cross-transplanted muscles, and was therefore present at less than 1 per cent, the detection limit of hydroxylamine-peptide analysis as performed here (Sandri *et al.* 1993).

7.3.3.4 *Gel densitometry*

In this work, gel electrophoresis was followed by densitometric scanning of gel slabs with a GS-300 Transmittance-Reflectance Scanning Densitometer (Hoefer Scientific Instruments) connected to an Apple Macintosh computer. Data were processed by the GS-370 Data System for the Hoefer GS-300 Scanning Densitometer (Macintosh version). Slabs were scanned first along the direction of electrophoretic migration and then along an axis perpendicular to this. Densitometric values were linear between 0.25 and 5 mg of protein after staining with Coomassie Blue (Sandri *et al.* 1992).

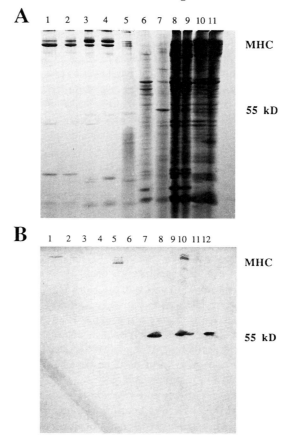

Fig. 7.6 SDS-PAGE and immunoblot analyses of hydroxylamine fragments of MHC$_{emb}$ from rat muscles in the early stages of regeneration. (A) Analysis on SDS 10 per cent PAGE with Coomassie Blue staining. Myosin heavy chains: 1, 5 μg of total protein from 2–day regenerating rat soleus; 2, 5 μg of total protein from 4–day regenerating rat soleus; 3, 6 μg of total protein from human transplanted latissimus dorsi muscle; 4, 6 μg of total protein from normal human pectoralis major muscle; 5, 2.5 μg of total protein from embryonic rat muscle. Hydroxylamine peptides of myosin from: 6, 5 μg of total protein from rat soleus; 7, 4 μg of total protein from embryonic rat muscle; 8, 25 μg of total protein from 2–day regenerating rat soleus; 9, 25 μg of total protein from 4–day regenerating rat soleus; 10, 25 μg of total protein from human transplanted latissimus dorsi muscle; 11, 25 μg of total protein from normal human pectoralis major muscle. (B) Immunoblot of a slab gel similar, but not identical, to that of (A) reacted with BF-G6, a monoclonal antibody to MHC$_{emb}$. Myosin heavy chains: 1, 4–day regenerating rat soleus; 2, 2–day regenerating rat soleus; 3, human transplanted latissimus dorsi muscle; 4, normal human pectoralis major muscle; 5, embryonic rat muscle. Hydroxylamine peptides of myosin from: 6, rat soleus; 7, embryonic rat muscle; 8, 2–day regenerating rat soleus; 9, 4–day regenerating rat soleus; 10, human transplanted latissimus dorsi muscle; 11, normal human pectoralis major muscle; 12, embryonic rat muscle. MHC, myosin heavy chains; 55 kD, hydroxylamine peptide of MHC$_{emb}$. The antibody to MHC$_{emb}$ recognizes the 55 kD peptide of MHC$_{emb}$ in both the embryonic and 4–day regenerating rat muscle.

7.3.4   Apoptosis

We have recently extended the analysis of cryosections to the demonstration of apoptotic myonuclei and DNA fragmentation in normal and dystrophin-deficient muscles of mice after spontaneous running (Podhorska-Okolow *et al.* 1995; Sandri *et al.* 1995). As this could shed fresh light on the pathogenesis of muscle dystrophy and exercise-induced muscle damage, the method will be described briefly here.

Cryostat sections are fixed in 10 per cent neutral buffered formalin for 10 min at room temperature. After the slides have been washed in phosphate-buffered saline, the sections are post-fixed in ethanol:acetic acid (2:1) for 5 min at room temperature. Proteins in the tissue sections are digested by applying proteinase K (20 mg/ml) for 15 min at room temperature. The slides are then washed in distilled water. Endogenous peroxidase activity is blocked by exposing the slides to 2 per cent hydrogen peroxide in phosphate-buffered saline for 5 min at room temperature. *In situ* end-labelling of fragmented DNA is performed by terminal deoxynucleotidyl transferase with digoxigenin-conjugated nucleotides, followed by immunodetection of incorporated nucleotide with the use of the ApopTag *in situ* Apoptosis Detection Kit (distributed by Oncor).

7.4   Conclusions

The assay of molecular markers from cryostat sections can be carried out simply and reliably. The values obtained with this approach for a larger series of normal EDL and soleus muscles in adult rats, expressed as mean $\pm$ SEM, are as follows:

> protein 21.9 $\pm$ 5.2 per cent;
> collagen 0.7 $\pm$ 0.4 per cent;
> myosin:actin ratio 2.1 $\pm$ 0.3.

These results are either similar or identical to those obtained with the use of standard methods on whole muscles (Reichel 1960).

The data obtained for the MHC profiles of normal whole EDL and soleus muscles in adult rats are shown in Table 7.3.

MHC_{emb} , a marker of ongoing muscle regeneration, was used as an example to show how the analysis can be extended when suitable antibodies are available for immunoblotting. In this way molecular markers of inflammation or of any other normal or pathological event can be included in a procedure that can be carried out on a few cryostat sections. Clearly the approach is not confined to skeletal muscle tissue.

Table 7.3 MHC profiles of EDL and soleus muscles of adult rats obtained with standard methods on whole muscles

	MHC profile (%)	
	EDL	Soleus
MHC1	0.5 ± 1.0	79.9 ± 9.0
MHC2A	—	20.1 ± 9.0
MHC2A + X	$36.1 \pm 4.3*$	†
MHC2B	63.3 ± 5.9	†

* When the bands are well separated, MHC2A:MHC2X = 0.5.
† MHC2X and MHC2B are not present.

Acknowledgements

This work was supported in part by funds from the Italian CNR to the Unit for Muscle Biology and Physiopathology, and the Italian Ministero dell'Università e della Ricerca Scientifica e Tecnologica (MURST) to U.C. The financial support of Telethon-Italy is gratefully acknowledged.

References

Ausoni, S., Gorza, L., Schiaffino, S., Gundersen, K., and Lømo, T. (1990). Expression of myosin heavy chain isoforms in stimulated fast and slow rat muscles. *Journal of Neuroscience*, **10**, 153–60.

Bär, A. and Pette, D. (1988). Three fast myosin heavy chains in adult rat skeletal muscle. *FEBS Letters*, **235**, 153–5.

Betto, R. (1993). Rivelazione delle proteine con argento-silver staining. In *Elettroforesi ed cromatografia di biopolimeri e loro frammenti* (ed. U. Carraro and L. Dalla Libera), pp. 136–7. Unipress, Padua.

Bornstein, P. and Balian, G. (1970). Cleavage at Asn-Gly bonds with hydroxylamine. *Methods in Enzymology*, **14**, 132–44.

Bradford, M. (1976). A rapid and sensitive method for the quantitation of microgram quantities of protein utilizing the principle of protein-dye binding. *Analytical Biochemistry*, **72**, 248–54.

Bronstein, I., Edwards, J.B., and Voyta, J.C. (1989). Novel chemiluminescent enzyme substrates. Applications to immunoassays. *Journal of Bioluminescence and Chemiluminescence*, **4**, 99–111.

Bronstein, I., Voyta, I.C., Murphy, O.J., Bresnick, L., and Kricka, L.J. (1992). Improved chemiluminescent Western blotting procedure. *BioTechniques*, **12**, 784–53.

Brooke, M.H. and Kaiser, K.K. (1970). Muscle fibre type: how many and what kind? *Archives of Neurology*, **23**, 369–79.

Cantini, M. and Carraro, U. (1993). Isolation of myoblasts from fast and slow regenerating muscles of adult rats. *Basic and Applied Myology*, **3**, 225–8.

Cantini, M., Fiorini, E., Catani, C., and Carraro, U. (1993). Differential expression of adult type MHC in satellite cell cultures from regenerating fast and slow rat muscles. *Cell Biology International*, **17**, 979–83.

Carraro, U. and Arpesella, G. (1991). Isomyosins in stimulated skeletal muscle. In *Cardiomyoplasty* (ed. A. Carpentier, J.-C. Chachques, and P. Grandjean), pp. 31–7. Futura Publications, Mount Kisko, New York.

Carraro, U. and Catani, C. (1983). A sensitive SDS-PAGE method separating myosin heavy chain isoforms of rat skeletal muscles reveals the heterogeneous nature of the embryonic myosin. *Biochemical and Biophysical Research Communications*, **116**, 793–802.

Carraro, U., Dalla Libera, L., and Catani, C. (1983). Myosin light and heavy chains in muscle regenerating in absence of the nerve. Transient appearance of the embryonic light chain. *Experimental Neurology*, **79**, 106–17.

Carraro, U., Morale, D., Mussini, I., Lucke, S., Cantini, M., Betto, R. *et al.* (1985). Chronic denervation of rat diaphragm: maintenance of fibre heterogeneity with associated increasing uniformity of myosin isoforms. *Journal of Cell Biology*, **100**, 161–74.

Carraro, U., Catani, C., Belluco, S., Cantini, M., and Marchioro, L. (1986). Slow-like electrostimulation switches on slow myosin in denervated fast muscle. *Experimental Neurology*, **94**, 537–53.

Carraro, U., Catani, C., Dell'Antone, P., Danieli-Betto, D., Arpesella, G., Parlapiano, M. *et al.* (1988). An experimental pumping chamber made with sheep latissimus dorsi: light microscopy and isomyosins. In *Sarcomeric and nonsarcomeric muscles: basic and applied research prospects for the 90's* (ed. U. Carraro and S. Salmons), pp. 459–70. Unipress, Padua.

Carraro, U., Catani, C., Degani, A., and Rizzi, C. (1990). Myosin expression in denervated fast and slow twitch muscles: fibre modulation and substitution. In *The dynamic state of muscle fibres* (ed. D. Pette), pp. 247–62. Walter de Gruyter, Berlin.

Danieli-Betto, D., Zerbato, E., and Betto, R. (1986). Type 1, 2A, 2B myosin heavy chain electrophoretic analysis of rat muscle fibres. *Biochemical and Biophysical Research Communications*, **138**, 981–7.

Dausse, E., Komajda, M., Fetler, L., Dubourg, O., Dufour, C., Carrier, L. *et al.* (1993). Familial hypertrophic cardiomyopathy. Microsatellite haplotyping and identification of a hotspot for mutations in the b-myosin heavy chains. *Journal of Clinical Investigation*, **92**, 2807–13.

Hall-Craggs, E.C.B. (1978). Ischemic muscle as a model of regeneration. *Experimental Neurology*, **60**, 393–9.

Hoffman, E.P., Brown, R.H., and Kunkel, M. (1987). Dystrophin: the protein product of the Duchenne muscular dystrophic locus. *Cell*, **51**, 919–28.

Huszar, G., Maiocco, J., and Naftolin, F. (1980). Monitoring of collagen and collagen fragments in chromatography of protein mixtures. *Analytical Biochemistry*, **105**, 424–9.

Jakubiec-Puka, A., Kulesza-Lipka, D., and Krajewski, K. (1981). The contractile apparatus of striated muscle in the course of atrophy and regeneration. I. Myosin and actin filaments in the denervated rat soleus. *Cell and Tissue Research*, **220**, 651–63.

Jakubiec-Puka, A., Kulesza-Lipka, D., and Kordowska, J. (1982). The contractile apparatus of striated muscle in course of atrophy and regeneration. II. Myosin and actin filaments in mature rat soleus muscle regenerating after reinnervation. *Cell and Tissue Research*, **227**, 641–50.

Jakubiec-Puka, A., Kordowska, J., Catani, C., and Carraro, U. (1990). Myosin heavy chain isoform composition in striated muscle after denervation and self-reinnervation. *European Journal of Biochemistry*, **193**, 623–8.

Jakubiec-Puka, A., Catani, C., and Carraro, U. (1992). Myosin heavy chain composition in striated muscle after tenotomy. *Biochemical Journal*, **282**, 237–42.

La Framboise, W.A., Daood, M.J., Guthrie, R.D., Moretti, P., Schiaffino, S., and Ontell, M. (1990). Electrophoretic separation and immunological identification of type 2X myosin heavy chain in rat skeletal muscle. *Biochimica Biophysica Acta*, **1035**, 109–12.

Laemmli, U.K. (1970). Cleavage of structural proteins during assembly of the head of bacteriophage. *Nature*, **227**, 680–5.

Lowry, O.H., Rosebrough, N.J., Farr, A.L., and Randall, R.J. (1951). Protein measurement with the Folin phenol reagent. *Journal of Biological Chemistry*, **193**, 265–75.

Mayne, C.N., Mokrusch, T., Jarvis, J.C., Gilroy, S.J., and Salmons, S. (1993). Stimulation-induced expression of slow muscle myosin in a fast muscle of the rat: evidence of an unrestricted adaptive capacity. *FEBS Letters*, **327**, 297–300.

McCapra, F. (1970). Chemiluminescence of organic compounds. *Pure and Applied Chemistry*, **26**, 611–19.

Merrill, R., Goldman Sedman, S.A., and Ebert, M.H. (1981). Ultra sensitive stain for proteins in polyacrylamide gels shows regional variation in cerebrospinal fluid proteins. *Science*, **211**, 1437–8.

Mussini, I., Favaro, G., and Carraro, U. (1987). Maturation, dystrophic changes and the continuous production of fibres in skeletal muscle regenerating in the absence of nerve. *Journal of Neuropathology and Experimental Neurology*, **46**, 315–31.

Pette, D. and Staron, R.S. (1990). Cellular and molecular diversity of mammalian skeletal muscle fibres. *Reviews of Physiology, Biochemistry and Pharmacology*, **116**, 1–76.

Pette, D. and Vrbová, G. (1992). Adaptation of mammalian skeletal muscle fibres to chronic electrical stimulation. *Reviews of Physiology, Biochemistry and Pharmacology*, **120**, 115–202.

Podhorska-Okolow, M., Sandri, M., Bruson, A., Carraro, U., Massimino, M.L., Arslan, P. *et al.* (1995). Apoptotic myonuclei appear in adult skeletal muscles of normal and *mdx* mice after mild exercise. *Basic and Applied Myology*, **5**, 87–90.

Reichel, H. (1960). Muskelphysiologie In *Lehrbuch der Physiologie* (ed. W. Trendelemburg and E. Schutz), pp. 1–276. Springer-Verlag, Berlin.

Rizzi, C. and Carraro, U. (1991). Electroendosmotic preparative gel electrophoresis and peptide mapping of slow and three fast myosin heavy chains. *Basic and Applied Myology*, **1**, 43–53.

Rossini, K., Rizzi, C., Sandri, M., Bruson, A., and Carraro, U. (1995). High-resolution sodium dodecyl sulphate-polyacrylamide gel electrophoresis and immunochemical identification of the 2X and embryonic myosin heavy chains in complex mixtures of isomyosins. *Electrophoresis*, **16**, 101–4.

Salmons, S. and Henriksson, J. (1981). The adaptive response of skeletal muscle to increased use. *Muscle and Nerve*, **4**, 94–105.

Sandri, M., Rizzi, C., Catani, C., and Carraro, U. (1992). Small and large scale preparative purification of myosin light and heavy chains by selective KDS precipitation of myosin subunits: yield by SDS PAGE and quantitative ortho-gonal densitometry. *Basic and Applied Myology*, **2**, 107–14.

Sandri, M., Rizzi, C., and Catani, C. (1993). Straightforward identification of MHCemb by hydroxylamine-peptide mapping. *Basic and Applied Myology*, **3**, 245–50.

Sandri, M., Carraro, U., Podhorska-Okolow, M., Rizzi, C., Arslan, P., Monti, M. *et al.* (1995). Apoptosis, DNA damage and ubiquitin expression in normal and mdx muscle fibers after exercise. *FEBS Letters,* **373**, 291–5.

Savolainen, J., Väänänen, K., Vihko, V., Puranen, J., and Takala, T.E.S. (1987). Effect of immobilization on collagen synthesis in rat skeletal muscle. *American Journal of Physiology*, **252** (*Regulatory and Integrative Comparative Physiology,* no. 21), R883–8.

Stephenson, L.W. (1993). Skeletal muscle-cardiac assist in the Americas (editorial). *Basic and Applied Myology*, **3**, 267–70.

Sugiura, T. and Murakami, N. (1990). Separation of myosin heavy chain isoforms in rat skeletal muscles by gradient sodium dodecyl sulphate-polyacrylamide gel electrophoresis. *Biomedical Research*, **11**, 87–91.

Szent-Gyorgyi, A. (1945). Studies on muscle. *Acta Physiologica Scandinavica*, **9**, 1–158.

Takahashi, H., Wada, M., and Katsuta, S. (1991). Expression of myosin heavy chain IId isoform in rat soleus muscle during hindlimb suspension. *Acta Physiologica Scandinavica*, **143**, 131–132.

Talmadge, R.J. and Roy, R.R. (1993). Electrophoretic separation of rat skeletal muscle myosin heavy chain isoforms. *Journal of Applied Physiology*, **75**, 2337–40.

Thomas, P.E. and Ranatunga, K.W. (1993). Factors affecting muscle fibre transfor-mation in cross-reinnervated muscle. *Muscle and Nerve*, **16**, 193–9.

Voyta, J.C., Edwards, B., and Bronstein, I. (1988). Ultrasensitive chemiluminescent detection of alkaline phosphatase activity. *Clinical Chemistry*, **34**, 1157–61.

Wilson, C.M. (1979). Studies and critique of amido black 10B, Coomassie blue R and Fast green FCF as a stain for protein after polyacrylamide gel electrophoresis. *Analytical Biochemistry*, **96**, 263–8.

Woessner, J.E. (1961). The determination of hydroxyproline in tissue and protein samples containing small proportions of this amino acid. *Archives of Biochem-istry and Biophysics*, **93**, 440–7.

8

Radioisotope uptake as a basis for investigating experimentally induced muscle damage

Valerie M. Cox, Jonathan C. Jarvis, Hazel Sutherland,
and Stanley Salmons

8.1   Introduction

In human volunteers or patients, muscle damage needs to be assessed in a minimally invasive way. Elsewhere in this volume reference is made to the assessment of damage in terms of pain (§§1.2.1 and 4.2), dysfunction (§1.2.2), range of joint motion (§§1.2.2, 4.2, and 9.2.1), and release into the circulation of muscle-specific proteins, such as creatine kinase, carbonic anhydrase III, and myoglobin (§§1.3.1, 1.3.2, and 4.2). Although adequate for some purposes, these measures do not localize the damage to a particular muscle, much less to a site within a muscle, and their relationship to the damaging process is not straightforward. More insight into the intracellular changes accompanying damage can be obtained from an analysis of biopsy samples (Chapters 7 and 10), but despite the obvious value of biopsy procedures—especially in the clinical setting (Chapter 10)—they have their limitations. They are invasive; they are not always well tolerated (Stroud 1993); they may not be permitted on ethical grounds; they cannot provide an indication of the distribution or extent of damage when it is focal in nature; and they themselves cause damage that has to be taken into account in any protocol that calls for serial sampling.

There is therefore considerable interest in other techniques that might be helpful in localizing and assessing muscle damage in human subjects. Computerized tomography and magnetic resonance imaging have been used to visualize fluid shifts and haemorrhage after eccentric exercise or strain injury (§§1.3.4 and 9.2.2; Fig. 9.1), and ^{31}P-magnetic resonance spectroscopy has been used to monitor the effects of damaging events on high-energy phosphate metabolism (§1.3.4). However, the technique that has been used most widely to date is gamma camera imaging of the uptake by damaged tissue of the radioactive tracer ^{99}technetium pyrophosphate (^{99}Tc^m-pyp).

This chapter is an attempt to address the question: what does $^{99}Tc^m$-pyp uptake actually represent? Studies are described in which low-level damage was induced in rabbit tibialis anterior (TA) and extensor digitorum longus (EDL) muscles by chronic electrical stimulation. We looked at the relationship between $^{99}Tc^m$-pyp uptake and damage, which was measured in a conventional histomorphometric way, and we made additional measurements of $^{99}Tc^m$-pyp uptake into the extracellular space and tissue blood volume. Based on the results we go on to describe a hierarchy of mechanisms whereby potentially damaging stimuli of escalating severity could produce apparent increases in $^{99}Tc^m$-pyp uptake without necessarily incurring permanent, irreversible muscle fibre damage.

8.2   Background to the use of ^{99}technetium

Originally the use of $^{99}Tc^m$-pyp was prompted by the need to visualize areas of increased bone turnover in disorders such as Paget's disease. $^{99}Tc^m$-pyp was chosen for its convenience of generation (from ^{99}Mo), short half-life (6 h), and characteristic emission spectrum. At first, the tripolyphosphate salt was used, but other phosphate salts were later developed (Subramanian and McAffee 1971; Krishnamurthy et al. 1975). When $^{99}Tc^m$-phosphates were used as bone-scanning agents, it was noted that the isotopes also entered other tissues in a variety of disease conditions (Brill 1981). It was this observation that led to the use of the isotope for demonstrating muscle damage. Bonte et al. (1974) were able to show increased uptake of $^{99}Tc^m$-pyp into experimentally infarcted canine hearts; similar observations were made in human patients suffering from myocardial infarction. Uptake of $^{99}Tc^m$-pyp was reported to increase in muscle trauma resulting from exercise, alcohol abuse, intramuscular injections, and radiotherapy (Brill 1981). Bellina et al. (1978) showed increased uptake in polymyositis, Steinert's disease (myotonic muscular dystrophy), and Duchenne muscular dystrophy. There were, however, no increases in patients with neurogenic myopathy, limb girdle myopathy, or myasthenia.

In the case of bone, $^{99}Tc^m$-pyp uptake is believed to be the result of adsorption on to hydroxyapatite crystals, although other mechanisms, such as hyperaemia or adsorption on to immature collagen, have been proposed (Rosenthal 1978). In other tissues the situation is even less clear. Some diseases involve an increase in tissue calcium; in many of these it has been possible to show an increase in $^{99}Tc^m$-pyp uptake (Dewanjee and Khan 1976; Reimer et al. 1976; Buja et al. 1977). This apparent association between calcium and $^{99}Tc^m$-pyp levels is not, however, a universal finding (Kaye et al. 1975; Silberstein and Bove 1979) so calcium binding is certainly not the only mechanism by which $^{99}Tc^m$-pyp becomes localized in damaged tissue. Despite the poor current state of understanding of the mechanisms involved, ^{99}Tc^m-pyp is used extensively to quantify damage under both experimental and clinical conditions.

Fast-twitch rabbit skeletal muscles, such as the TA and EDL muscles, show an adaptive response to long-term electrical stimulation, as a result of which the fibres become extremely resistant to fatigue (Salmons and Henriksson 1981; Pette 1984). The therapeutic applications of this phenomenon are attracting increasing interest (see §§11.1.1 and 11.1.2). However, a small proportion of fibres in chronically stimulated muscles may show histological evidence of damage, and this is a matter for some concern in a clinical context (§5.1 and Chapter 11).

The original goal of the work described here was to quantify this damage in its early stages by the use of ^{99}Tc^m-pyp. Although preliminary experiments with a gamma probe and a gamma camera showed that there was an increase in ^{99}Tc^m-pyp uptake in limbs stimulated at either 2.5 or 10 Hz, these methods of detection were not capable of localizing the increase to a specific tissue. We therefore used radioactive counting to measure the amount of isotope present in samples of muscle.

8.3   Methodological details

8.3.1   Operative procedures

New Zealand White rabbits of both sexes, with body weights within the range 2.5–4.0 kg, were prepared for aseptic surgery with a pre-anaesthetic medication of diazepam (Valium® Roche Products Ltd, 5 mg/kg) and atropine sulphate (Bimeda, 3 mg/kg) delivered subcutaneously, followed after 20 minutes by an intramuscular injection of Hypnorm® (Janssen Pharmaceuticals: fluanisone, 10 mg/ml and fentanyl citrate, 0.315 mg/ml; 0.3 ml/kg). The stimulators used in these experiments generated a continuous train of pulses at either 2.5 or 10 Hz, and were switched on and off remotely via an optical link (Jarvis and Salmons 1981). The devices were implanted under the skin of the left flank, and the leads were conducted subcutaneously to the left hindlimb, where the electrodes were sutured close to the common peroneal nerve. Animals were allowed to recover from the operation for at least 2 weeks before the stimulation was commenced. Isotopes were administered 3 days after the onset of stimulation.

8.3.2   Administration of radioisotopes

^{99}Tc^m-pyp was obtained from the Radiopharmacy Department of the Royal Liverpool University Hospital. It was prepared fresh daily at an approximate dosage of 200 MBq per rabbit and injected as a single bolus via the marginal ear vein. The rabbits were killed 3 h later by overdose of pentobarbitone sodium. The protocol was established in a pilot study as the one that produced most radioactivity in the muscles.

^{51}Cr-EDTA was obtained from Amersham International at a specific activity of 50 μCi/ml. As this substance is cleared rapidly via the kidneys, these organs had to be isolated surgically before the isotope was administered. This was achieved by anaesthetizing the animal and ligating the vessels entering the hilum of each kidney. ^{51}Cr-EDTA was then injected via the marginal ear vein. The animals remained under anaesthesia for a further hour and were then killed by overdose.

^{51}Cr-labelled erythrocytes (^{51}Cr-RBC) were prepared by a protocol recommended by the Radiopharmacy Department of the Royal Liverpool University Hospital. Approximately 2 ml of blood were removed via the marginal ear vein of the rabbit, and transferred immediately to a heparinized tube. The sample was then spun at 1500 g for 7 min. After the supernatant was discarded, 0.4 ml of ^{51}Cr (Amersham, 50 μCi/ml) was added to the tube and vortex-mixed. The sample was spun again at 7000 g for 7 min, and the supernatant again discarded. The pellet was dissolved in 2 ml of saline, mixed and spun for the last time at 1500 g for 7 min. The pellet was dissolved in 2 ml of saline, and kept at room temperature for up to 1 h before being reintroduced into the rabbit. Each rabbit was killed by anaesthetic overdose 1 h after being injected with approximately 2 ml of its own labelled red blood cells. A blood sample was removed by cardiac puncture before the animals were killed.

8.3.3 Tissue sampling

At termination, the TA and EDL muscles of both hindlimbs were quickly removed. In some of the experiments other tissues, such as bone and skin, were also taken.

The muscles were cut transversely into 10–mm slices, and the 3 middle sections were used for histological analysis and radioactive counting. The blood samples were kept at room temperature for 2 h and were then spun down for 10 min in a Beckman Microfuge. The serum was removed and stored at $-20°C$.

8.3.4 Counting radioactivity

Tissue samples were counted in an LKB-Wallac 1270 Rack Gammacounter. For all animals, samples of both blood and serum were counted. In experiments involving administration of ^{51}Cr-RBC, a sample of the labelled cell preparation was also counted.

Since $^{99}Tc^m$ has a half-life of only 6 h, the radioactive counts obtained for this isotope had to be corrected for the time taken to count all the samples. All the values were calculated as cpm/g of tissue, and corrected for the different specific activities of the serum sample (not the blood sample) taken from each animal at termination.

All data is given as mean $\pm$ SEM. Significance was tested by the paired Student's t-test for left-right comparisons, and by the unpaired t-test for comparison of different treatments.

8.3.5 Histology

Cryostat sections of 10 μm thickness were air-dried for 1 h before being stained conventionally with haematoxylin and eosin. Quantification of damage was carried out by the method of Lexell *et al.* (1992). In brief, a transparent overlay carrying a grid of 1 mm squares was placed over the coverslip of the muscle sections. Each square was further subdivided into 10×10 cells by an eyepiece graticule. The nature of the tissue at each of the 121 intersections of the graticule grid was noted as a normal fibre, a damaged fibre, or interfibrillar tissue. The total number of fibres, normal and damaged, that fell within the area of the graticule grid was then determined. For each of the TA and EDL sections, between 8 and 12 one-millimetre squares were evaluated in this way. The percentage of intersect points occupied by the three categories of tissue provided a measure of their relative contribution to the total tissue volume, and the number of fibres gave the percentage affected by damage.

8.4   Results

8.4.1 $^{99}Tc^m$-pyp levels in blood

Only 75 per cent of the $^{99}Tc^m$-pyp counts in blood could be found in the serum; the rest must have been associated with blood cells, either within cells or bound to surface molecules.

8.4.2 $^{99}Tc^m$-pyp levels in stimulated muscle

The column chart of Fig. 8.1 shows the results of stimulating the left TA and EDL muscles continuously for 3 days at 2.5 Hz ($n = 5$) or 10 Hz ($n = 10$), with muscles from the unstimulated contralateral limb and from sham-operated animals ($n = 2$) included for comparison. For both patterns of stimulation, muscles from the stimulated side contained significantly more

^{99}Tc^m-pyp than their contralateral counterparts (10 Hz: $p<0.01$; 2.5 Hz: $p<0.02$). The level of ^{99}Tc^m-pyp was significantly greater in muscles stimulated at 10 Hz than in those stimulated at 2.5 Hz ($p<0.001$). The levels of ^{99}Tc^m-pyp in the unstimulated contralateral muscles did not differ significantly for the two stimulation protocols; nor did they differ from levels in the left or right limbs of the sham-operated animals. Thus the increased isotope uptake in the stimulated muscles could not be attributed to the operative procedures alone.

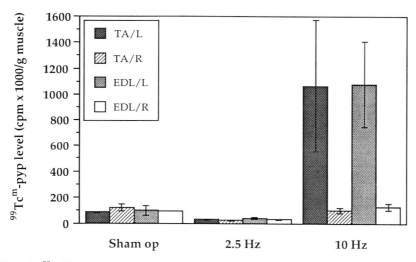

Fig. 8.1 ^{99}Tc^m-pyp levels in muscles of the left hindlimb that received stimulation at 2.5 or 10 Hz (TA/L, EDL/L) and their contralateral controls in the right hindlimb (TA/R, EDL/R). Data from the corresponding muscles in animals that underwent sham operations are also included. Error bars denote mean ± SEM.

There was no significant difference between the values obtained for the left and right soleus muscles, neither of which had received stimulation. However, both of these muscles had levels of ^{99}Tc^m-pyp that were significantly higher than in the muscles stimulated at 2.5 Hz (see below). Predominantly slow muscles, such as the soleus, are known to have proportionally higher extracellular fluid (ECF) volumes than fast muscles. This suggested that at least some of the increases we were seeing in the stimulated muscles might be due to altered tissue fluid volumes, rather than isotope uptake. To test this hypothesis we examined the uptake of ^{51}Cr-EDTA, a substance that is restricted to the extracellular space (Poole-Wilson and Cameron 1975).

8.4.3 ^{51}Cr-EDTA levels in muscle

^{51}Cr-EDTA levels were expressed as a percentage of serum activity. This value is equivalent to the percentage of the tissue volume that consists of fluid of the same specific activity as the serum. Provided that the ^{51}Cr-EDTA is excluded from cells, this percentage represents the volume fraction of ECF in the tissue. Muscles that were stimulated at 2.5 Hz had a mean ECF of 11–15 per cent (Fig. 8.2). The values for the (unstimulated) soleus muscles were higher: about 20 per cent. These values were within the 10–20 per cent range found by previous investigators, who also observed differences in ECF between slow and fast muscles (Sréter and Woo 1963; Poole-Wilson and Cameron 1975). The muscles that were stimulated at 10 Hz had a much larger ECF than those that were stimulated at 2.5 Hz, but even in these muscles the fluid volumes were not significantly larger than those in the soleus muscles (Fig. 8.2(B)).

For both TA and EDL muscles, there were significant differences in ^{51}Cr-EDTA levels between the stimulated and unstimulated sides, whether stimulation was at 10 Hz ($n = 2$) or 2.5 Hz ($n = 5$). In the 10 Hz-stimulated animals, however, even the unstimulated EDL muscles, and both soleus muscles, had levels of the isotope that were significantly above control values. Thus stimulation at 10 Hz does produce some generalized oedema and/or increase in blood volume in the stimulated limb, and there is an indication from this data that these changes may extend to the contralateral limb.

8.4.4 Apparent ^{99}Tc^m-pyp volumes

The serum ^{99}Tc^m-pyp levels were used in the same way as the serum ^{51}Cr-EDTA levels, to predict the percentage of tissue volume that would be occupied by fluid of the same specific activity of ^{99}Tc^m-pyp as the serum. These results are referred to as percentage ^{99}Tc^m-pyp volume. If the ^{99}Tc^m-pyp had been freely distributed throughout the extracellular fluid in the muscles, but not taken up into the muscles fibres, the percentage ^{99}Tc^m-pyp volume would be equal to the ECF.

The results of this analysis are given in Table 8.1, which also includes data from the soleus muscles alluded to earlier. The percentage ^{99}Tc^m-pyp volumes were considerably lower than the values for ECF determined by ^{51}Cr-EDTA, and the differences were highly significant ($p < 0.004$) for all the muscles. These results suggested that the ^{99}Tc^m-pyp was not free to distribute throughout the extracellular space.

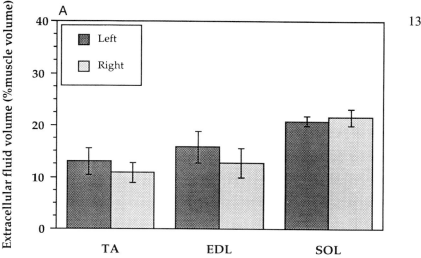

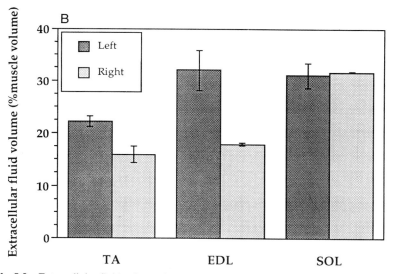

Fig. 8.2 Extracellular fluid volumes in muscles that received stimulation at (A) 2.5 Hz or (B) 10 Hz (left) and their contralateral controls (right). Data is included for the (unstimulated) soleus muscles of the same limbs.

8.4.5 ^{51}Cr-labelled erythrocytes

To estimate the volume of the vascular bed in muscle samples we used ^{51}Cr-labelled erythrocytes (^{51}Cr-RBC), since these are retained within the blood vessels (Baker 1963). Blood volumes were measured in four animals in which muscles were stimulated at 10 Hz, and the results of these experiments are shown in Fig. 8.3, expressed as percentages of muscle volume. As expected, the largest blood volumes were found in the soleus muscles. Comparison of the stimulated TA and EDL muscles with their unstimulated contralateral counterparts failed to reveal any significant difference between the blood volume fractions

Table 8.1 The effects of stimulation at 2.5 or 10 Hz on the volumes of compartments in the TA and EDL muscles, and the unstimulated soleus muscles of the corresponding sides*

	TA/L	TA/R	EDL/L	EDL/R	SOL/L	SOL/R
2.5 Hz						
^{51}Cr-EDTA volume (ECF) ($n=5$)	13.0 ± 2.6	10.9 ± 1.9	15.8 ± 3.1	12.7 ± 2.8	20.9 ± 0.9	21.6 ± 1.6
Apparent ^{99}Tc^{m}-pyp volume ($n=5$)	2.3 ± 0.3	1.8 ± 0.2	3.1 ± 0.4	2.2 ± 0.4	3.2 ± 0.3	2.9 ± 0.4
10 Hz						
^{51}Cr-EDTA volume (ECF) ($n=2$)	22.2 ± 1.0	15.9 ± 1.6	32.0 ± 3.9	17.9 ± 0.2	31.0 ± 2.4	31.6 ± 0.1
Apparent ^{99}Tc^{m}-pyp volume ($n=10$)	79.2 ± 37.8	7.2 ± 1.4	80.0 ± 24.5	9.4 ± 1.8	—	—
Serum volume ($n=4$)	4.7 ± 1.0	4.3 ± 0.9	5.0 ± 1.0	4.7 ± 1.3	7.8 ± 2.3	6.0 ± 1.6

* Values are mean ± SEM.

estimated by ^{51}Cr-RBC. Thus the differences we observed (Fig. 8.1) between stimulated and non-stimulated muscles in their uptake of [99]Tc[m]-pyp could not be attributed to changes in the blood volume of these tissues.

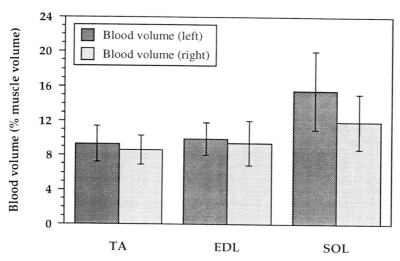

Fig. 8.3 Blood volumes in muscles that received stimulation at 10 Hz (left) and their contralateral controls (right). Data is included for the (unstimulated) soleus muscles of the same limbs.

In both the TA and EDL muscles, the blood space occupied about 10 per cent of muscle volume. In the rabbit, the haematocrit is close to 50 per cent, so the serum volume is half the blood volume. The blood volumes shown in Fig. 8.3 have been converted to serum volumes in Table 8.1 to permit comparison with percentage [99]Tc[m]-pyp volumes. The percentage [99]Tc[m]-pyp volumes were significantly larger than the values for serum volume in the EDL muscles stimulated at 10 Hz ($p < 0.05$).

8.4.6 Morphometric assessment of damage

Sections taken from muscles from the unstimulated side contained few or no damaged fibres; this was also true of TA muscles that had received stimulation at 2.5 Hz. All other muscles contained a small percentage of fibres that had mononuclear cells within their cytoplasm and were clearly undergoing degeneration. These degenerating fibres appeared to be distributed randomly throughout the muscle sections. The amount of damage seen in muscles stimulated at the two different frequencies is given in Table 8.2. There was a significant overall correlation between [99]Tc[m]-pyp uptake and damage ($r = 0.68$; $p < 0.0001$), but the relationship varied considerably from one muscle to another (Fig. 8.4).

Table 8.2 Morphometric data from muscle samples

Stimulation frequency	TA/L	TA/R	EDL/L	EDL/R
Percentage volume of interfibrillar tissue				
2.5 Hz ($n=5$)	7.5 ± 2.5	4.5 ± 0.5	8.75 ± 1.44*	5.1 ± 1.0
10 Hz ($n=10$)	7.32 ± 0.56	6.14 ± 0.74	7.99 ± 1.11	7.49 ± 2.49
Percentage volume of damaged fibres				
2.5 Hz ($n=5$)	0.15 ± 0.05	0 ± 0	0.6 ± 0.21*	0 ± 0
10 Hz ($n=10$)	0.43 ± 0.2*	0 ± 0	1.32 ± 0.53*	0.014 ± 0.014
Percentage of fibres showing damage				
2.5 Hz ($n=5$)	0.28 ± 0.02	0.08 ± 0.08	1.125 ± 0.72	0 ± 0
10 Hz ($n=10$)	0.715 ± 0.33*	0.01 ± 0.01	1.62 ± 0.58*	0.036 ± 0.024

*$p < 0.05$ for experimental versus contralateral muscles.

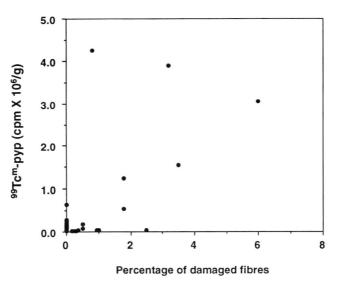

Fig. 8.4 $^{99}Tc^m$-pyp uptake plotted against the percentage of fibres showing damage in stimulated (2.5 and 10 Hz) and control rabbit TA and EDL muscles.

8.4.7 Isotope uptake into skin and bone

The various volumes described above were also calculated for samples of skin and bone tissue from rabbits that received the two stimulation patterns. The results are shown in Fig. 5(A), (B). There were no significant differences between the counts obtained from the left and right limbs for any of the

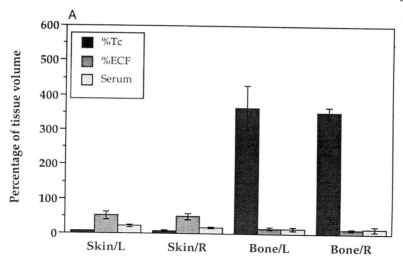

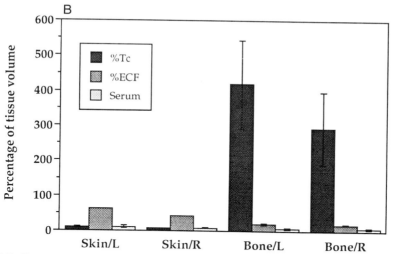

Fig. 8.5 Percentage volumes of fluid compartments in skin and bone tissue from the left (/L) and right (/R) hindlimbs of rabbits in which the left common peroneal nerve was stimulated at (A) 2.5 Hz or (B) 10 Hz.

volumes, in either skin or bone. In skin, the values for ECF were as high as 60 per cent. This was unlikely to represent the true ECF, and it may be that ^{51}Cr-EDTA is localized inside cells, or bound to matrix components in this tissue. Skin also showed higher levels of percentage ^{99}Tc^m-pyp volume than muscle, with values of about 8 per cent. However, there was no significant difference between the serum volume and the percentage $^{99}Tc^m$-pyp volume, and $^{99}Tc^m$-pyp therefore appears to be confined to the serum in skin. In contrast, the levels of $^{99}Tc^m$-pyp in bone far exceeded those that could be accounted for in terms of serum volume, and it is evident that the isotope is not confined to the intravascular space in this tissue.

8.5   Discussion

In this study we have sought clues to the mechanisms underlying the uptake of $^{99}Tc^m$-pyp by damaged muscle. Since the short half-life and long path-length of the γ-emission from $^{99}Tc^m$ make it unsuitable for localizing directly by high-resolution autoradiography, we used measurements of muscle and serum activity, together with determinations of extracellular space and tissue blood volume, to trace the passage of the isotope between tissue compartments.

We used counts of $^{99}Tc^m$-pyp in muscle and serum to compute the volume that $^{99}Tc^m$-pyp would occupy if it were concentrated at serum levels in the muscle samples. In control muscles, and in muscles that were stimulated at 2.5 Hz, these percentage $^{99}Tc^m$-pyp volumes were lower than the volume fractions occupied by extracellular fluid, estimated by the ^{51}Cr-EDTA technique. This suggested either that the $^{99}Tc^m$-pyp was not freely distributed within the extracellular fluid compartment, or that the ^{51}Cr-EDTA was not confined to the extracellular space, and was therefore overestimating its volume. The values for percentage $^{99}Tc^m$-pyp volume in the muscles that received stimulation at 2.5 Hz were about 3 per cent. This is much lower than the level of 10–20 per cent usually quoted for tissue fluid volume in muscle (Sréter and Woo 1963; Poole-Wilson and Cameron 1975). It is unlikely, therefore, that the percentage $^{99}Tc^m$-pyp volume represented the true extracellular fluid volume. It was, in fact, similar to published figures for serum volume in muscles, and this suggested the possibility that the $^{99}Tc^m$-pyp may have been confined to the blood vessels.

The ^{51}Cr-RBC technique was therefore used to determine the percentage of volume occupied by the vascular bed in these muscles. For most of the muscle samples in this study, the volume occupied by serum was very similar to the volume that $^{99}Tc^m$-pyp would have occupied at serum concentration. We therefore conclude that in muscles $^{99}Tc^m$ is normally confined to the serum.

The data for skin is likewise consistent with $^{99}Tc^m$-pyp being present only in the serum. In bone, however, $^{99}Tc^m$-pyp is present at more than three times the serum level. In normal bone, therefore, the isotope must be able to leave the capillaries. Why $^{99}Tc^m$-pyp should be confined to the serum in some tissues and not in others is not clear. However, the permeability of capillaries is known to vary at different sites in the body—contrast, for example, the highly impermeable capillaries of the cerebral circulation with the extremely permeable capillaries of the renal glomeruli.

There was a clear difference between stimulated and non-stimulated muscles in their uptake of $^{99}Tc^m$-pyp. However, in terms of the volumes occupied by the vascular bed, the stimulated TA and EDL muscles did not differ from their unstimulated contralateral counterparts. The increased uptake of $^{99}Tc^m$-pyp in stimulated muscles could not therefore be explained as a mere consequence of changes in blood volume in the tissues. We deduce that under these conditions, $^{99}Tc^m$-pyp was able to leave the confines of the serum. The permeabilization of capillaries underlying such an event could be one of the earliest responses to a potentially damaging stimulus. It has been suggested that $^{99}Tc^m$-pyp uptake is greater in muscles from patients with inflammatory disorders (such as polymyositis) than non-inflammatory disorders (such as Steinert's disease) (Bellina *et al.* 1978). These observations, too, may be understood in terms of an increase in the permeability of capillaries associated with the inflammatory process.

In which tissue compartment was the extravascular $^{99}Tc^m$-pyp localized? In muscles stimulated at 2.5 Hz, the $^{99}Tc^m$-pyp uptake could be accounted for by the presence of the isotope in an enlarged extracellular space. In these samples, the increases in the uptake of both $^{99}Tc^m$-pyp and ^{51}Cr-EDTA could have been the consequence of changes in ECF towards values normally found in slow muscles, rather than the result of increased uptake into cells. However, most of the muscles stimulated at 10 Hz had concentrated $^{99}Tc^m$-pyp above the serum level, and this could occur only by uptake into cells.

It could be postulated that $^{99}Tc^m$-pyp was taken up only by damaged fibres. However, this and other studies with the same rabbit model (see, for example, Chapter 5) have shown that less than 5 per cent of the fibres in a muscle are damaged after 3 days of stimulation. If the uptake of $^{99}Tc^m$-pyp were confined to these fibres, then the concentration of $^{99}Tc^m$-pyp within them would have to be 40 times the serum level to account for the average concentration seen in the muscle. 'Hot spots' in rat bone can concentrate $^{99}Tc^m$-pyp to 11 times the normal bone level (Kaye *et al.* 1975), which is itself 3 times the serum level; therefore bone can probably concentrate $^{99}Tc^m$-pyp some 30–fold. However, two-thirds of bone consists of inorganic matrix with a strong affinity for $^{99}Tc^m$-pyp, and it is unlikely that muscle cytoplasm can bind as much $^{99}Tc^m$-pyp as an equivalent mass of calcium-rich bone matrix. It therefore seems improbable that $^{99}Tc^m$-pyp is concentrated only in fibres exhibiting frank histological damage. The argument that $^{99}Tc^m$-pyp is confined to a non-muscle cell population may be similarly dismissed, since such cells make up only a small percentage of the muscle bulk.

There are two other possible explanations. The first is that $^{99}Tc^m$-pyp is taken up by cells of normal histological appearance that are, nevertheless, damaged in a *reversible* way. This could occur in the majority of the fibres in the muscle during the early stages of stimulation, and $^{99}Tc^m$-pyp could be taken up to some extent by all of these fibres. After 9 days of stimulation, when histological damage appears to be maximal (Lexell *et al.* 1992), the fibres infiltrated with mononuclear cells would be the comparatively few that had been damaged beyond their capacity for self-repair. The second possible explanation is that all muscle fibres can concentrate $^{99}Tc^m$-pyp once they are exposed to it in the extracellular fluid. In an undamaged muscle, $^{99}Tc^m$-pyp would remain within the serum; permeabilization of capillaries in the damaged muscle would release the isotope into the extracellular space. $^{99}Tc^m$-pyp is known to bind strongly to calcium ions, which muscle cells take up from the extracellular fluid and store within the sarcoplasmic reticulum and, under some conditions, mitochondria. In this way $^{99}Tc^m$-pyp levels twice those in serum could readily be attained within intact muscle fibres. Although such a mechanism could be operative when the viability of the tissue is only mildly challenged, permeabilization of the sarcolemma would also have to take place to account for the other common accompaniment of muscle damage: the release into the circulation of muscle-specific proteins such as creatine kinase, carbonic anhydrase III, and myoglobin (§§1.3.1, 1.3.2, and 4.2).

Is there a threshold level of damage, above which large amounts of $^{99}Tc^m$-pyp are taken up, or is there a direct proportionality between damage and $^{99}Tc^m$-pyp levels? Although there was a significant correlation between the $^{99}Tc^m$-pyp level and the amount of damage in muscles, several muscles showed either low isotope uptake in the presence of damage, or a slightly higher isotope uptake in the absence of damage (Fig. 8.4). Uptake of $^{99}Tc^m$-pyp is not, therefore, very predictable at low levels of muscle damage. On the other hand, none of the muscles that had concentrated $^{99}Tc^m$-pyp above the serum level were undamaged, so the technique may be a more reliable indicator of damage at these higher levels of uptake. Possibly these two conditions correspond to the two explanations given above: a low and variable level of isotope uptake from a pool of $^{99}Tc^m$-pyp released into the extracellular fluid, and a higher uptake by muscle fibres rendered more permeable by the damaging stimulus. More refined morphological methods will be needed to test this hypothesis directly.

8.6   Conclusions

We conclude that increased $^{99}Tc^m$-pyp uptake can take place through a hierarchy of mechanisms. In the simplest case, an apparent increase in uptake can be the result of a change in the volume of the vascular bed within which the isotope is normally confined. A potentially damaging stimulus can bring about a larger uptake as a result of permeabilization of capillaries, accompanied by oedema and accumulation of $^{99}Tc^m$-pyp in the expanded extracellular space. This creates the conditions in which concentration of the isotope into the muscle fibres can occur, but such a process is not necessarily dependent upon permeabilization of the sarcolemma. Even when the insult to the muscle is sufficient for permeabilization of the sarcolemma to occur, the fibres involved need not always sustain permanent, irreversible damage. Clearly, an increased uptake of $^{99}Tc^m$-pyp by muscle in clinical or experimental situations needs to be interpreted with some caution.

Acknowledgements

We wish to acknowledge the value to us of a pilot study carried out in this laboratory by Miss J. Williams and Dr C.N. Mayne. We thank Dr J. Lexell for help and advice, particularly with regard to the morphometric estimation of damage. We are grateful to the staff of the Radiopharmacy Department of the Royal Liverpool University Hospital for technical assistance and advice on the use of the radioisotopes.

References

Baker, C.H. (1963). Cr^{51}-labelled red cells, I^{131}-fibrinogen, and T-1824 dilution spaces. *American Journal of Physiology,* **204**, 176–80.

Bellina, C.R., Bianchi, R., Bombardieri, S., Ferri, C., Mariani, G., Muratorio, A. *et al.* (1978). Quantitative evaluation of $^{99}Tc^m$-pyrophosphate muscle uptake in patients with inflammatory and noninflammatory muscle disease. *Journal of Nuclear Medicine and Allied Science,* **22**, 89–96.

Bonte, F.J., Parkey, R.W., Graham, K.D., Moore, R.T.J., and Stokely, E.M. (1974). New method for radionuclide imaging of MI. *Radiology,* **110**, 473–4.

Brill, D.R. (1981). Radionuclide imaging of non-neoplastic soft tissue disorders. *Seminars in Nuclear Medicine,* **XI**, 277–88.

Buja, L.M., Tofe, A.J., Kulkarni, P.V., Mukherjee, A., Parkey R.W., Francis, M.D. *et al.* (1977). Sites and mechanisms of localisation of technetium^{99m}-phosphorus radio-pharmaceuticals in acute MI and other tissues. *Journal of Clinical Investigation,* **60**, 724–40.

Dewanjee, M.K. and Khan, P.C. (1976). Mechanism of location of $^{99}Tc^m$-labelled pyrophosphate and tetracycline in infarcted myocardium. *Journal of Nuclear Medicine,* **17**, 639–46.

Jarvis, J.C. and Salmons, S. (1981). A family of neuromuscular stimulators with optical transcutaneous control. *Journal of Medical Engineering and Technology,* **15**, 53–7.

Kaye, M., Silverton, S., and Rosenthall, L. (1975). Technetium-99–m-pyrophosphate: studies in vivo and in vitro. *Journal of Nuclear Medicine,* **16**, 40–5.

Krishnamurthy, G.T., Huebotter, R.J., Walsh, C.F., Taylor, J.R., Kehr, M.D., Tubis, M. *et al.* (1975). Kinetics of $^{99}Tc^m$-labelled pyrophosphate and polyphosphate in man. *Journal of Nuclear Medicine,* **16**, 109–15.

Lexell, J., Jarvis, J.C., Downham, D.Y., and Salmons, S. (1992). Quantitative morphology of stimulation-induced damage in rabbit fast-twitch muscles. *Cell and Tissue Research,* **269**, 195–204.

Pette, D. (1984). Activity induced fast to slow transitions in mammalian muscle *Medicine and Science in Sport and Exercise,* **16**, 517–28.

Poole-Wilson, P.A., and Cameron, I.R. (1975). ECS, intracellular pH, and electrolytes of cardiac and skeletal muscle. *American Journal of Physiology,* **229**, 1299–304.

Reimer, K.A., Marton, K., Schumacher, B.L., Henkin, R.E., Quinn, J.L., and Jennings, R.B. (1976). Cardiac localisation of ^{99m}Tc-pyrophosphate after temporary or permanent coronary occlusion in dogs. *American Journal of Pathology,* **82**, 23A-4A.

Rosenthal, L. (1978). ^{99m}Tc methylene diphosphonate concentration in soft tissue malignant fibrous histocytoma. *Clinical Nuclear Medicine,* **3**, 58–61.

Salmons, S. and Henriksson, J. (1981). The adaptive response of skeletal muscle to increased use. *Muscle and Nerve,* **4**, 94–105.

Siegel, J., Osmand, A.P., Wilson, M.F., and Gexurtz, H. (1975). Interactions of C reactive protein with the complement system II C reactive protein mediated consumption of complement by poly L lysine polymers and other polycations. *Journal of Experimental Medicine,* **142**, 709–21.

Silberstein, E.B. and Bove, K. (1979). Visualisation of alcohol induced rhabdomyolysis: a correlative radiotracer, histochemistry and em study. *Journal of Nuclear Medicine,* **20**, 127–9.

Sréter, F.A. and Woo, G. (1963). Cell water, sodium, and potassium in red and white mammalian muscles. *American Journal of Physiology,* **205**, 1290–4.

Stroud, M. (1993). *Shadows on the wasteland.* Jonathan Cape, London.

Subramanian, G. and McAfee, J.G. (1971). A new complex of ^{99m}Tc for skeletal imaging. *Radiology,* **99**, 192–6.

9

Muscle strain injuries: biomechanical and structural studies

Thomas M. Best, Carl T. Hasselman, and William E. Garrett, Jr.

9.1 Muscle strain injuries

Acute injuries to skeletal muscle—including contusions, lacerations, ruptures, ischaemia, and stretch injuries—are common and often lead to significant pain, disability, and time lost from work and athletic pursuits. The clinical significance of stretch-induced muscle injuries, or 'strains', is well known to those who treat occupational and sports-related injuries. This type of injury is usually said to be the most common injury in sports, and has been estimated to involve 30 per cent of a clinical practice in sports medicine (Ryan 1969; Kreji and Koch 1979; Glick 1980; Peterson and Renström 1986). The morbidity associated with muscle strain injury has an effect both on the individual and on society as a whole, part of which is the economic burden imposed by time lost, cost of rehabilitation, and in some cases disability due to chronic pain.

In spite of the frequency of acute muscle strain injuries, and their potentially disabling effects, our understanding of their pathophysiology, treatment, and outcome is limited. One of the possible reasons for the relative lack of clinical and basic research in muscle injury has been the lack of a clear association with a particular clinical specialty. Muscle injury is so common that it is treated by most primary care practitioners, yet recovery is usually excellent and mortality virtually non-existent, so the topic has not gained a large research following. Muscle injuries rarely call for surgical intervention and have not, therefore, become a particular interest of orthopaedic surgeons. The injuries have a higher profile in Sports and Occupational Medicine, because of their impact on the ability to work or to play. The lack of basic and clinical research into these injuries has attracted comment from National Health agencies (Garrett and Tidball 1987).

Non-contact or indirect injuries to skeletal muscle include delayed onset muscle soreness (DOMS), partial strain injury, and complete muscle rupture. The three conditions have the common feature that they are more prone to occur with eccentric exercise (Glick 1980; Zarins and Ciullo 1983; Peterson and Renström 1986). DOMS is characterized by generalized muscle soreness and weakness which is maximal 1–2 days after eccentric exercise (see, for example, Fig. 4.1) and is not usually associated with long-term morbidity (Asmussen 1956; Stauber 1989). In contrast, partial strain injury and complete muscle rupture present as more focal pain and weakness noted during exercise, and often carry significant morbidity (Stauber 1989). This is the common 'muscle pull' or 'strain' that is experienced as an acute painful injury by an athlete during exercise. In this chapter we will focus on acute muscle strain injuries, including partial injury and total muscle rupture. Clinical observations will be taken together with data from animal experiments in an effort to develop a consensus view of the pathophysiology and mechanism of injury, as well as methods of treatment and prevention.

9.2   Clinical studies of muscle strain injury

9.2.1   Mechanism of injury

Although there is a large amount of clinical data on injury to skeletal muscle, the exact mechanism of muscle strain injury remains unknown. It is widely accepted in clinical circles that muscle strain injuries occur in response to forcible stretch of a muscle, either passively or—more often—while the muscle is activated, the condition known as 'eccentric contraction' (Kreji and Koch 1979; Radin *et al.* 1979; Glick 1980; Zarins and Ciullo 1983; Peterson and Renström 1986). Eccentric contractions may make a muscle more prone to injury because the large forces produced by the contractile element are added to the extrinsic forces stretching the muscle. The muscles most susceptible to injury are the 'two-joint' or biarticular muscles; these cross two or more joints and are therefore subject to stretch at more than one joint (Brewer 1960; see also §5.9). Other muscles that are frequently injured are those that limit the range of motion of a joint. For example, hamstring muscles can limit knee extension when the hip is flexed; similarly, gastrocnemius can limit ankle dorsiflexion when the knee is extended. Susceptible muscles also include those that are often called upon to function in an eccentric manner in the control and regulation of movement. Much of the muscle action involved in running or sprinting is eccentric (Cavagna 1977; Cavagna *et al.* 1977; Bosco *et al.* 1982, 1987). For example, the hamstring muscles act not so much to flex the knee as to decelerate the lower extremity during knee extension. Similarly, the quadriceps group acts as much to prevent knee flexion as to power knee extension in running (Cavagna *et al.* 1977; Mann and Hagy 1980; Inman *et al.* 1981). Muscle strain injuries occur most often in sprinters or 'speed athletes' and are common in sports and positions that require rapid changes in velocity, such as American football, basketball, rugby, soccer, and sprinting (Peterson and Renström 1986). Typically, the affected muscles will tend to have a high percentage of type

2, or fast-twitch, muscle fibres (Garrett *et al.* 1984*a*). The predisposition to injury of these muscles may be a reflection of the intrinsic properties of type 2 fibres, but it should be noted that muscles with this predominant composition are, in general, located more superficially and often cross more than one joint.

9.2.2 Structural changes with muscle strain injury

There are few clinical data on the exact nature of changes within the muscle following strain injury. The injury can be partial or complete, depending on whether the muscle-tendon unit is grossly disrupted (Kreji and Koch 1979). Complete tears are characterized by muscle asymmetry at rest when comparison is made to the contralateral contour. With contraction, the injured muscle will demonstrate a bulge towards the side that is still attached to bone. Direct muscle injury or contusion produces injury at the place of contact. Until recently, however, the location of pathological changes in a muscle following strain injury has been less well defined. The vulnerable site in strain injury appears to be near to the muscle-tendon junction (MTJ) or near to the tendon-bone junction (Garrett *et al.* 1988). The extent of the MTJ in certain muscles is often a surprise to clinicians. The hamstring muscles, for example, have an extended MTJ in the posterior thigh. The extent of the proximal tendon and MTJs of biceps femoris and semimembranosus is well over half the total length of these muscles.

Although surgical exploration of muscle strain injuries is not commonly performed, there are a number of references to surgical findings in the literature. These studies have confirmed tears near to the MTJ in:

- the medial head of gastrocnemius (often incorrectly called 'plantaris rupture') (McMaster 1933; Durig *et al.* 1977–78);
- rectus femoris (Rask and Lattig 1972; Hasselman *et al.* 1995*a*);
- triceps brachii (Bach *et al.* 1987);
- adductor longus (Symeonides 1972);
- pectoralis major (Peterson and Renström 1986);
- semimembranosus muscle (Garrett *et al.* 1988).

More recently, high-resolution imaging studies have localized hamstring injuries to the region of the MTJ (Garrett *et al.* 1989). CT scans revealed areas of hypodensity consistent with oedema at the site of injury in all patients. In another study, 50 athletes with acute muscle strain injuries were imaged by CT or MRI within 96 h of injury (Speer *et al.* 1993). With MRI, T_2-weighted images were most effective for identifying free fluid shifts, and for differentiating these areas from injured tissue. T_1-weighted images were most appropriate as an adjunct to these, specifically for identifying haemorrhage (Fig. 9.1). It was interesting to observe that, within a given synergistic muscle group, certain muscles were more prone to injury than others. For example, adductor longus was the most frequently injured muscle of the adductor group and rectus femoris was the most frequently injured muscle of the quadriceps group. The reasons for this are currently unknown. In all cases, injury occurred at the MTJ, and involved fluid collections at the injury site extending along the epimysium. Muscle tissue remote from the MTJ showed extensive injury and changes consistent with oedema and inflammation.

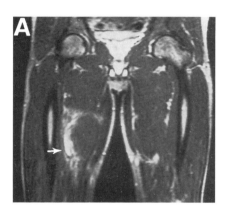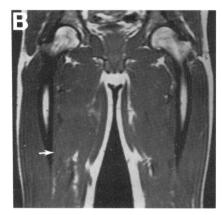

Fig. 9.1   (A) T_2-weighted coronal image (repetition time, 2000 ms; echo time, 70 ms) of a strain in the adductor longus muscle. The arrow identifies oedema, inflammation, and possible haemorrhage. (B) T_2-weighted coronal image (repetition time, 600 ms; echo time, 20 ms) of the same injury as in (A). The arrow indicates the site of injury. This imaging modality cannot differentiate adequately between injured and uninjured muscle tissue. (Adapted from Speer *et al.* 1993.)

Although bleeding often occurs following muscle injury, it usually takes one or more days before subcutaneous ecchymosis is detectable. Subcutaneous collections of blood show that the bleeding is not confined to the muscle proper; rather it escapes through the epimysium and fascia to the subcutaneous space. In other cases a haematoma can collect between the muscle tissue and the surrounding fascial compartment, and can be demonstrated by ultrasonography (Fornage *et al.* 1983; Harcke *et al.* 1988). This finding contrasts with what is seen in a direct muscular contusion, where bleeding is often confined to the muscle substance (Rothwell 1982; Rooser 1988).

9.2.3   Treatment and prevention

The treatment strategy for muscle strain injury has varied considerably (Kreji and Koch 1979; Peterson and Renström 1986) and is often based on what is successful in clinical practice rather than on basic science. The immediate treatment usually involves rest, application of ice, and compression of the affected soft tissue and extremity. Additional treatment modalities typically include physical therapy to improve the range of motion, strengthening exercises (Glick 1980), and medication. Medication has included anaesthetics, analgesics, muscle relaxants, and anti-inflammatory (steroidal and non-steroidal) agents. Administration of non-steroidal anti-inflammatory agents is usually initiated as soon after the injury as possible and used for a short period (5–7 days) since these drugs may interfere with subsequent tissue repair and remodelling. Surgical intervention has been advocated by some in cases of complete dissociation of the muscle-tendon unit (Reed and Woodward 1963; Miller 1977; Kreji and Koch 1979; Glick 1980; O'Donoghue 1984; Chammout and Skinner 1986). A period of enforced rest—involving, for example, a brace or cast—is sometimes recommended in cases of severe injury (Smodlaka 1980; Abbe 1995).

Precautions are often taken in an effort to avoid muscle strain injury. Most athletes routinely practise stretching mainly because it is thought to prevent muscle injury (Beaulieu 1981; Solveborn 1983; Wiktorsson-Moller *et al.* 1983). Adequate warm-up is also cited as a way of preventing muscle injury (Kreji and Koch 1979). Training programmes that employed adequate stretching and warm-up seemed to help in reducing the incidence of muscle injuries (Ekstrand *et al.* 1983), but these programmes incorporated a number of variables and the effects of individual factors were not established.

The adoption of precautionary measures, such as stretching and warm-up, is often coupled with avoidance of additional risk factors, such as fatigue (Kreji and Koch 1979) and a history of previous injury (Peterson and Renström 1986). Although these are widely held to be important risk factors for muscle injury, there have been few solid clinical or laboratory studies to support this belief.

9.3 Experimental data on acute muscle strain injury

Skeletal muscle is subjected to a wide range of loading conditions *in vivo* and it is difficult to reproduce these conditions experimentally. One of the first studies of muscle strain injury showed that rupture did not occur in normal tendon when the muscle-tendon unit was stretched to failure (McMaster 1933); failure occurred at the tendon-bone junction, the MTJ, or within the muscle. Previous studies from this laboratory focused on recovery in rabbit hindlimb muscles following muscle laceration and repair (Garrett *et al.* 1984*b*). We have since adapted this model to the study of acute muscle strain injuries (Fig. 9.2). In initial experiments it was shown that activation of normal muscle by nerve stimulation alone caused no disruption, complete or incomplete (Garrett *et al.* 1988): force diminished and failure of excitation occurred, but no disruption was seen. In order to produce gross or micro-scopic muscle injury, stretch of the muscle was required. The forces produced at the time of failure, even without muscle activation, were several times the maximum isometric force produced by the muscle itself (Taylor *et al.* 1993). We interpreted this data to suggest that the passive component of the total muscle force might be as important as the active component in strain injuries. Since these initial experiments were completed, we have conducted a number of studies to investigate various aspects of muscle strain injury.

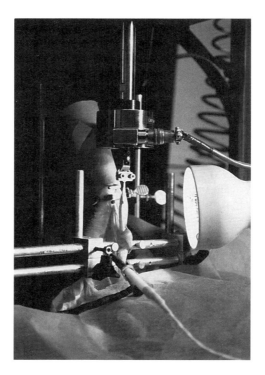

Fig. 9.2 Experimental set-up, showing the distal tendon of a rabbit EDL muscle clamped to the Instron piston and aligned with the axis of motion of the piston. A heating lamp and warm normal saline are used to maintain muscle temperature at 38 ± 0.1°C. A stimulating electrode is placed in contact with the common peroneal nerve.

9.3.1 Site of injury

Clinical studies have shown that the MTJ is the most common site of damage in acute muscle strain injuries (Rask and Lattig 1972; Symeonides 1972; Durig *et al.* 1977–78; Miller 1977; Peterson and Renström 1986; Bach *et al.* 1987; Garrett *et al.* 1989; Speer *et al.* 1993; Hasselman *et al.* 1995*b*).

Laboratory studies, in which a variety of test preparations were used, have demonstrated injury that occurred for the most part at or near the MTJ (Garrett and Tidball 1987; Garrett *et al.* 1987; Reddy *et al.* 1993; Taylor *et al.* 1993; Best *et al.* 1995; Hasselman *et al.* 1995*b*). Why the MTJ should be especially susceptible to injury is not easy to understand. It would not have been predicted from a maximum stress theory of failure, as the MTJ experiences the same loads as the distal tendon but distributed over a larger cross-sectional area. Recent data suggests that the MTJ may be predisposed to injury by large strains that occur locally (Best *et al.* 1995).

9.3.2 Viscoelastic properties of muscle

Biological tissues, including skeletal muscle, are inhomogeneous, anisotropic, and demonstrate non-linear viscoelastic behaviour (Abbott and Lowy 1956; McElhaney 1964; Noyes *et al.* 1974; Peterson and Woo 1986; Woo *et al.* 1990; Best *et al.* 1995). The stress in the tissue at any given time depends not only on the strain at that moment, but also on the history of applied strain and the strain rate. Stress relaxation and creep are important characteristics of viscoelastic materials (Noyes *et al.* 1974). We have studied these phenomena in our laboratory. Initial experiments, in which the muscle was cycled repetitively to a fixed displacement, revealed a decrease in peak force with successive stretches (Taylor *et al.* 1989). Eventually, consecutive stretches showed no difference in peak force and energy absorption, a phenomenon known as 'mechanical stabilization', or 'preconditioning'. In our model, the creep and stress relaxation responses were found to be independent of reflex effects or other influences mediated by the central nervous system (Taylor *et al.* 1989). The independence of stress relaxation and creep changes in the muscle-tendon unit from nervous activity has been confirmed in a further study (Cole *et al.* 1990).

Our interest in muscle viscoelasticity has led to studies in which this behaviour was modelled by the quasi-linear theory of viscoelasticity (Best *et al.* 1994). This theory, first proposed for the study of the mechanical response of rubber, was adopted by Dr. Y.C. Fung to describe the viscoelastic properties of biological tissues (Fung 1967). The measured load response, $F(t)$, of a tissue to an applied deformation history, $\delta(t)$, can be expressed as the convolution integral of a reduced relaxation function, $G_r(t)$, and the elastic response, $F^e(\delta)$, of the tissue by

$$F(t) = \int_0^t G_r(t - \tau) \frac{\partial F^e(\delta)}{\partial \delta} \frac{\partial \delta}{\partial \tau} d\tau$$

where $F(t) = 0$ and $\delta(t) = 0$, for $t < 0$. When muscles were subjected to constant deformation experimentally, all muscle-tendon units showed similar relaxation behaviour: a monotonically decreasing function that approached asymptotically a non-zero equilibrium value, G_{inf} (Fig. 9.3). It is important to note the variable rate of load relaxation in Fig. 9.3. Initially, the load decay was extremely rapid; thereafter, the load decayed at a much slower rate. This observed relaxation behaviour cannot be modelled by a standard lumped parameter viscoelastic model with a single dominant long-term time constant. Furthermore, this type of relaxation behaviour is not well described by traditional Hill-type models. The force-time response of the muscle to constant-velocity perturbations was compared with the response produced by the model (Fig. 9.4; Best *et al.* 1994). Testing was carried out with 22 muscles from seven rabbits. The model performed well, with an overall correlation of 0.93 ± 0.01 between the observed data and the response predicted from the model for the 132 tests performed. The results of these experiments may have significant clinical implications. The rapid load relaxation of the muscle-tendon unit in response to constant deformation may help to explain why athletes stretch and warm up prior to competition. These experiments also show that the rate and amount of stretch can affect the observed response of the muscle. It may be important to account for these factors when stretching protocols are being developed. Finally, the time dependency of the relaxation behaviour seen in muscle probably cannot be explained by a reflexly mediated response.

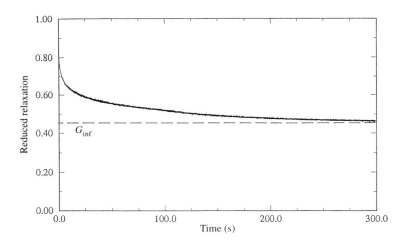

Fig. 9.3 A typical measured relaxation response, normalized to peak load, showing an asymptotic approach to a non-zero equilibrium value, G_{inf}.

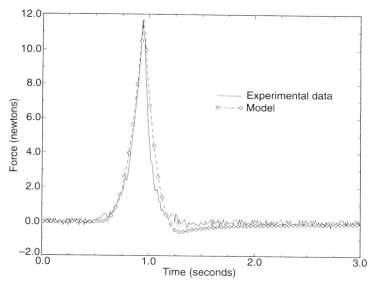

Fig. 9.4 Force-time response to a 10 mm/s triangular waveform test, showing the excellent correlation between the experimental data and the response predicted by the model ($r = 0.98$).

9.3.3 Passive stretch

The effects of passive stretch were evaluated with five muscles that had different fibre architectures (Garrett *et al.* 1988). Muscles were stretched from either the proximal or distal tendon without preconditioning or muscle activation. Strain rates were 10, 100, and 1000 mm/min. Injury consistently occurred near the MTJ, usually distally. A small and variable amount of muscle fibre, usually 1–2 mm, was left attached to the tendon. These experiments demonstrated that, within the range of strain rates used, disruption occurred in a predictable way near the MTJ for all of the muscles, regardless of their architectural features. Classical physiological studies of muscle have demonstrated that the active force-generating capability of the muscle is proportional to its cross-sectional area, and shortening ability is proportional to the muscle fibre length. These concepts have been applied to human muscle performance (Wickiewicz *et al.* 1983). We felt that the biomechanical response of a muscle to stretch might also be related to fibre length. There was, however, no evidence of this; the amount of strain ranged widely from 75 to 225 per cent of the resting fibre length.

9.3.4 Active stretch

In clinical practice it is believed that muscle injuries occur during powerful eccentric contractions, and experiments were devised to evaluate these conditions. Muscles were stretched to failure under three types of motor nerve activation: tetanic stimulation; submaximal stimulation; and no stimulation (Garrett *et al.* 1987). The results were somewhat unexpected. Muscle stimulation did not change the site of failure (the MTJ). Total deformation to failure did not differ between the three groups. Force at failure was 15 per cent higher in stimulated muscles. However, the energy absorbed was approximately 100 per cent higher in muscles that were stretched to failure while activated. These data confirm the importance of considering muscles as energy absorbers. The passive components of stretched muscle are able to absorb energy, but the capacity for absorbing energy is greatly increased by active muscle contraction. The above concept may help to explain how muscle can prevent injury both to itself and to associated joint structures. The ability of the muscle to absorb energy has two components. The passive component is not dependent on muscle activation and is a property of the connective tissue elements within the muscle, including the extracellular matrix. The active component lies within the contractile element of the muscle. From the above experiments, the contractile element can double the ability of the muscle to absorb energy (Fig. 9.5). Indeed, at small muscle displacements, most of the energy absorption is due to the active component (Lieber *et al.* 1991; Lieber and Fridén 1993). Therefore, conditions that reduce the contractile ability of muscle may also diminish its ability to absorb energy. This is consistent with the commonly held view that muscle fatigue and muscle weakness are factors that predispose to muscle injury (Kreji and Koch 1979).

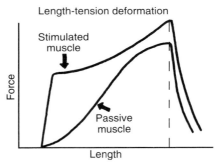

Fig. 9.5 Schematic diagram comparing force-length relationships of passive and stimulated muscle stretched to failure. Note that the two muscles fail at the same length. At small displacements, energy absorption is much greater for the active muscle. (Adapted from Garrett *et al.* 1987.)

9.3.5   Nondisruptive (partial) injury

So far we have considered the biomechanics of muscle stretched to failure, or complete muscle disruption. Most strain injuries to muscle do not result in total rupture and dissociation of the MTJ unit, but rather in incomplete or partial disruption. These partial strain injuries have also been investigated experimentally. Non-rupture injuries were created by stretching a muscle passively to 80 per cent of the force needed to produce failure in the contralateral muscle. Histological studies show that injuries that are non-disruptive at the whole muscle level do cause disruption of a small number of muscle fibres near to the MTJ (Fig. 9.6; Nikolaou *et al.* 1987; Reddy *et al.* 1993). The fibres do not rupture at the actual junction between fibre and tendon; rather the tear occurs within the fibre a short distance from the tendon. Our experiments have rarely produced disruption near the middle of a muscle fibre. In the acute phase the injuries were marked by disruption and some haemorrhage within the muscle. By 1 to 2 days a pronounced inflammatory reaction was observed, with invasion by inflammatory cells and oedema. By 7 days the inflammatory reaction was being replaced by an increase in fibrous tissue near the region of the injury. Although some regenerating muscle fibres were present, normal histology was not restored and scar tissue persisted (Fig. 9.7). The functional recovery of the muscle was determined by testing physiologically the maximal force produced in response to nerve stimulation (Taylor *et al.* 1993). Immediately after the injury, the muscle generated 70 per cent of the force produced by non-injured control muscle. By 24 hours, force production was only 50 per cent of normal. This trend then reversed, and by 7 days force production was 90 per cent of normal, so recovery of the contractile apparatus was fairly rapid.

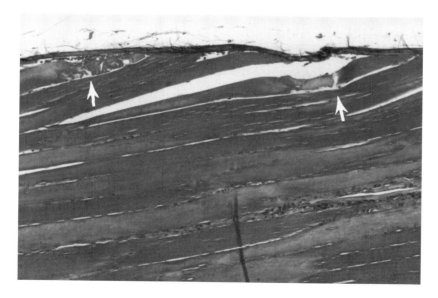

Fig. 9.6 Light micrograph of a longitudinal section of muscle; magnification, 10 ×. Arrows point to disrupted fibres near to the distal muscle-tendon junction.

These results were subsequently confirmed in a study designed to evaluate the influence of non-steroidal anti-inflammatory drugs (NSAIDs) on healing of muscle injuries (Obremsky *et al*. 1988).

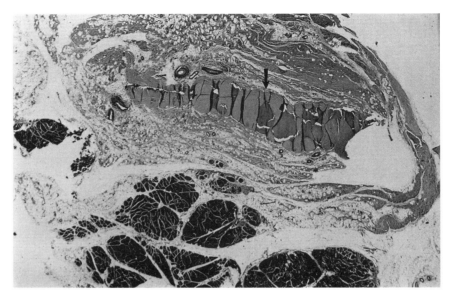

Fig. 9.7   Microscopic view of an axial section taken through scar tissue. Arrowhead points to the tendon. Note the encasement of the tendon in vascular loose connective tissue. Muscle fibres can be seen arising from this loose connective tissue at the bottom of the micrograph. Section stained with Masson's trichrome; magnification, 10 × .

The tensile strength of muscle following a nondisruptive strain injury has also been evaluated (Obremsky *et al*. 1988). Seven days post-injury, tensile strength was 77 per cent of that of the contralateral non-injured control muscles, whereas active force-generating ability approached 90 per cent of control. These studies suggest that tensile strength may be a suitable indicator of the susceptibility of muscle to injury. Because of this loss of tensile strength, a previous injury may predispose a muscle to a second injury.

Subsequent studies have involved the use of electron microscopy to evaluate further the immediate structural response at the MTJ to acute injury (Reddy *et al*. 1993). Sixty minutes post-injury, sarcomeres closest to the site of fibre rupture were hypercontracted (Fig. 9.8(A)). There was a progressive increase in sarcomere length away from the rupture site, normal lengths being reached by 300–500 μm (Fig. 9.8(B)). Six hours after injury, necrosis was present in the region of transition between hypercontracted and normal sarcomeres. These findings suggest that an intracellular barrier effectively restricts the response to acute injury to less than 500 μm from the initial site of rupture.

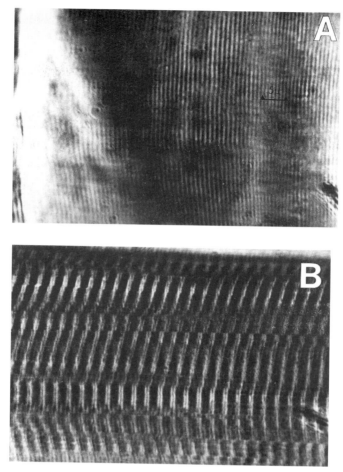

Fig. 9.8 Electron micrographs of unstained material showing a disrupted fibre near to the muscle-tendon junction. Magnification, $100 \times$. (A) High power view of a 100 μm segment at the site of rupture. The average sarcomere length (measured as the distance between adjacent A-bands) is 0.86 μm, indicative of severe contraction. (B) The same fibre 1000 μm from the site of rupture. The sarcomeres appear uniform, with an average length of 2.61 μm. Z-discs are more clearly visible in this segment. (Adapted from Reddy *et al.* 1993.)

We have tested the hypothesis that injury is a graded phenomenon (Table 9.1). We showed that, at the smallest forces necessary to produce injury, the contractile ability of the muscle was impaired, the amplitude of the electromyogram (EMG) was diminished, and a variable degree of injury occurred near the MTJ (Hasselman *et al.* 1995*b*). At higher levels of applied force, contractile ability and EMG amplitude continued to decline, and both the MTJ and the distal muscle belly were injured. Forces more than 80 per cent of those that produced failure in the control muscle resulted in changes in the passive tensile properties, as well as diminished contractile ability and EMG amplitude. Histological examination confirmed that there was disruption at the MTJ, and damage in the distal muscle belly and in the connective tissue. It therefore appears that there is a continuum for acute muscle strain injury. In another study, an eccentric stretch was applied to the muscle to create a partial strain injury. The ability of the muscle to generate force after the injury was tested by both nerve and direct muscle stimulation, neuromuscular transmission being blocked for the latter with vecuronium. The results showed that, although most of the reduction in contractile ability was due to muscle fibre injury, some was due to injury to intramuscular nerve branches or neuromuscular junctions (Pedowitz *et al.* 1995).

Table 9.1 Schematic summary to illustrate the continuum of injury observed as the amount of tissue displacement increases. Note that in all cases disruption of muscle fibres precedes disruption of connective tissue

	Percentage strain*			
	13.3	14.6	17.8	23.2
Maximum isometric contractile force	Unchanged	Diminished 20%	Diminished 50%	Diminished 80%
Failure properties	Unchanged	Unchanged	Unchanged	Altered
Histology: muscle belly	Normal	Oedema; normal muscle fibres	Oedema and bleeding; inflammatory cells; few disrupted fibres	Oedema and bleeding; inflammatory cells; many disrupted fibres; connective tissue disruption
Histology: muscle-tendon junction	Normal	Oedema; inflammatory cells; focal fibre disruption	Oedema and bleeding; inflammatory cells; moderate fibre disruption	Oedema and bleeding: inflammatory cells; massive fibre disruption; connective tissue disruption

* Percentage strain $= 100(L-L_0)/L_0$; L, final length; L_0, initial length.

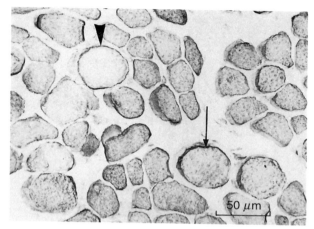

Fig. 10.3 Hyaline fibres in Duchenne dystrophy, one showing normal desmin immunolabelling (arrow) and the other showing reduced immunolabelling (arrowhead).

et al. 1990), and titin is lost before desmin in Duchenne dystrophy (Cullen *et al.* 1992). Studies of desmin with two antibodies have revealed that in many human myopathies the loss of one epitope in some damaged fibres is accompanied by increased, granular staining for a second epitope (Figs 10.4 and 10.5), suggesting that partial degradation leads to collapse of the normal desmin lattice (Helliwell *et al.* 1989). Progressive damage to both desmin and dystrophin occurs after exposure to 2,4–dinitrophenol *in vitro* (Helliwell *et al.* 1994*b*) but not after calcium ionophore, suggesting that the processes responsible for damage are different. The preferential loss of labelling by one of two antibodies to titin has been observed in some fibres in Duchenne dystrophy (Cullen *et al.* 1992).

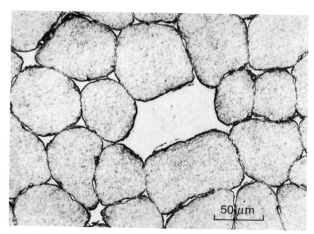

Fig. 10.4 Loss of desmin immunolabelling in a necrotic fibre (antibody DE-R-11, Dakopatts).

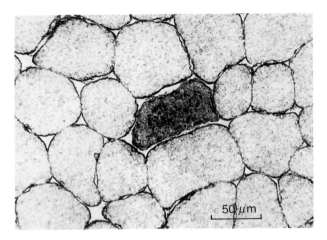

Fig. 10.5   Section serial to Fig. 10.3, showing increased desmin immunolabelling in the necrotic fibre (antibody D33, Dakopatts).

10.4   Muscle fibre regeneration

As phagocytes remove the debris from necrotic fibres, satellite cells are activated and may undergo mitosis. Myofibrillar synthesis starts within 1–2 days and the cells are then termed myoblasts. The number of mononucleate myoblasts peaks 2–3 days after injury, when they fuse to form elongated myotubes with multiple nuclei and a cytoplasm that is basophilic owing to the high content of RNA. As the cells enlarge they lose their basophilia but retain a central chain of nuclei for several weeks or months before the nuclei migrate to the periphery of the fibre (Carlson 1973; Allbrook 1981). Incomplete fusion of myoblasts into myotubes leads to irregular splits or branches in the fibres, and these are a common feature of dystrophic muscle (Isaacs *et al.* 1973; Bradley 1979; Head *et al.* 1990; Snow 1990).

Regenerating fibres have a histochemical profile similar to that of immature muscle (Engel 1979): they are not differentiated by ATPase staining, and they show a high content of oxidative enzymes and cytoplasmic staining for RNA and alkaline phosphatase (Engel and Cunningham 1970). The increased turnover of cytoplasmic proteins during regeneration is reflected in increased activity of lysosomal hydrolase (Engel 1973).

10.4.1   Immunohistochemistry of regeneration

The immaturity of regenerating fibres is indicated by weak immunoreactivity for membrane-associated proteins such as spectrin and dystrophin (Fig.10. 6; Appleyard *et al.* 1984; Vater *et al.* 1992) and by the expression of proteins

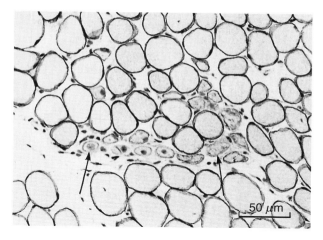

Fig. 10.6 β-spectrin immunolabelling is weak in regenerating fibres in Duchenne dystrophy (arrows).

usually associated with developing muscle such as fetal myosin (Fitzsimmons and Hoh 1981; Sartore *et al.* 1982; Schiaffino *et al.* 1986), neural cell adhesion molecule (NCAM; Walsh and Moore 1985; Figarella-Branger *et al.* 1990), the intermediate filament nestin (Sjöberg *et al.* 1994), laminin A (Mundegar *et al.* 1995), utrophin (Helliwell *et al.* 1992c), desmin (Fig. 10.7; Thornell *et al.* 1980, 1983; Helliwell 1988; Helliwell *et al.* 1989), and vimentin (Bornemann and Schmalbruch 1991, 1993). Although immunohistochemical labelling of the basement membrane and the sarcolemma will allow satellite cells to be identified (Zhang and McLennan 1994), and M-cadherin is expressed at the surface of satellite cells and myoblasts (Bornemann and Schmalbruch 1994),

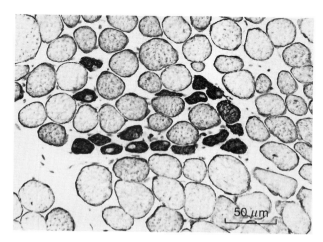

Fig. 10.7 Increased desmin immunolabelling in regenerating fibres in Duchenne dystrophy.

there are no completely specific markers for quiescent satellite cells. Activated satellite cells show the presence of early muscle-specific genes (MyoD and myogenin), and immunolabelling is strong for NCAM, desmin, and vimentin (Helliwell 1988; Figarella-Branger *et al.* 1990; Bornemann and Schmalbruch 1991; Yablonka-Reuveni and Rivera 1994). In human muscle after experimental muscle damage and in *mdx* mice, the expression of desmin and (in myoblasts) fetal myosin is a valuable aid in recognizing the earliest stages of regeneration (Fig. 10.7). Some care must be taken in the interpretation of labelling in small fibres, as desmin and utrophin immunolabelling are also increased in neurogenic atrophy (Helliwell *et al.* 1989; Helliwell 1993). Vimentin expression is often difficult to interpret because it is present in endothelial cells and fibroblasts, but expression in regenerating fibres is lost 2 weeks after experimental damage (Bornemann and Schmalbruch 1991).

10.4.2 Satellite cells and the control of regeneration

The key element in regeneration is the satellite cell (Mauro 1961), which is mononucleate, devoid of cytoplasmic myofibrils, and situated outside the plasma membrane—but within the basement membrane—of a mature muscle fibre. Satellite cells are the source of new nuclei in growing muscle (Moss and Leblond 1971), participate in regeneration (Church *et al.* 1966; Lipton and Schultz 1979; Schultz 1990), and are activated after sublethal damage resulting from eccentric contraction (Darr and Schultz 1987). They are also activated during the hypertrophy that results from ablation of synergists or administration of clenbuterol (Snow 1990; Maltin and Delday 1992).

Satellite cells in normal muscle are maintained in a quiescent state by contact with the plasmalemma, possibly by electrical activity and/or the production of growth inhibitory factors (Bischoff 1990). Satellite cells may be activated by growth factors, proteases, or other substances released from damaged cells (Bischoff 1986; Snow 1990; Grounds 1991). Once activated, their differentiation is inhibited by fibroblast growth factor (FGF) and transforming growth factor-β (TGF-β), so these factors allow myoblasts to continue to proliferate in culture (Gospodarowicz *et al.* 1976; Massagué *et al.* 1986, 1991; Yablonka-Reuveni and Rivera 1994). Insulin-like growth factor (IGF-I) may affect the expression of transferrin receptor on regenerating fibres (Jennische 1989) and stimulates protein synthesis, cell proliferation, and differentiation (Allen and Rankin 1990; Ewton and Florini 1990). Muscle growth may also be affected by other cytokines, including leukaemia inhibitory factor, interleukin 6, transforming growth factor-a, and platelet-derived growth factor (Austin and Burgess 1991; Austin *et al.* 1992). Growth factors and hormones probably interact with oncogene proteins, since the effects of dexamethasone, FGF, and TGF-β on differentiation are very similar to those of increased *ras* expression. IGF-I may act via the nuclear oncoprotein, c-*fos* (Schneider and Olson 1988; Florini and Magri 1989). In addition, c-*myc* suppresses differentiation and may balance the differentiating effects of other genes such as MyoD (Miner and Wold 1991; Weintraub *et al.* 1991).

10.5   Sublethal injury

Even when damage to muscle fibres is insufficient to cause necrosis, histological and ultrastructural changes in fibre structure may still be evident. Such sublethal damage has been identified under experimental conditions and in a range of human myopathies and neuropathies.

10.5.1   Experimental muscle damage

Tenotomy causes temporary myofibrillar breakdown at the centre of fibres, and this is associated with non-lysosomal proteolysis (Karpati *et al.* 1972). A similar pattern of myofibrillar breakdown occurs in central core disease and in childhood dermatomyositis (Carpenter *et al.* 1976; Karpati and Carpenter 1984). Exocytosis of membranous whorls and fibre splitting may be manifestations of injury in experimental chloroquine myopathy (MacDonald and Engel 1970) and are occasionally seen in polymyositis.

10.5.2   Denervation and reinnervation

In denervated muscles, there is often a loss of oxidative enzyme activity in the central region of muscle fibres accompanied by an increase in activity around the central zone; this produces the appearance known as 'target fibres'. These changes are reflected in increased immunolabelling for desmin and dystrophin in the central region (Vita *et al.* 1994*b*). There is increased immunolabelling for the calcium-binding protein, parvalbumin, during denervation and reinnervation (Olive and Ferrer 1994). Cytoplasmic labelling for desmin is increased in atrophic fibres during neurogenic atrophy (Helliwell *et al.* 1989) and the expression of utrophin and NCAM on the fibre surface is increased (Figarella-Branger *et al.* 1990; Helliwell 1993).

10.5.3   Acid maltase deficiency

In acid maltase deficiency, autophagic vacuoles appear which contain glycogen and cytoplasmic debris. Dystrophin, vinculin, and spectrin are present at the edges of the vacuoles and desmin appears to show increased labelling in this region, possibly corresponding to modification of the membrane of the phagocytic vacuole before exocytosis (Vita *et al.* 1994*a*).

10.5.4 Congenital myopathies

In congenital myopathies there are structural abnormalities of specific components of muscle fibres. These changes are often reflected in abnormal patterns of immunolabelling for a variety of proteins, either as the primary defect or, more frequently, as a secondary manifestation of other changes in fibre architecture. Immunolabeling for desmin appears to provide a particularly sensitive method of detecting changes in other cytoskeletal components, as the normal, uniformly distributed perimyofibrillar lattice is affected by changes to most other components of the myofibrils. In nemaline myopathy, the rods are formed by aggregates of α-actinin and actin with desmin present around the edges (Yamaguchi *et al.* 1978; Jockusch *et al.* 1980; Thornell *et al.* 1983; Hashimoto *et al.* 1989). In central core disease, there is decreased desmin labelling in the core region and increased labelling in other parts of the fibres, although whether the redistribution of desmin is primary or secondary is not known (Vita *et al.* 1994*b*).

Bundles of cytoplasmic filaments are occasionally observed in a wide range of muscle diseases (Goebel and Bornemann 1993), but discrete cytoplasmic inclusion bodies that contain desmin filaments are a characteristic feature of a group of hereditary distal myopathies, which often occur in association with cardiac conduction defects or a restrictive cardiomyopathy (Edström *et al.* 1980; Porte *et al.* 1980; Osborn and Goebel 1983). Filament assembly may be influenced by abnormal phosphorylation of desmin in some cases (Rappaport *et al.* 1988), while in other cases the desmin filaments are associated with dystrophin (Figs 10.8 and 10.9; Prelle *et al.* 1992; Helliwell *et al.* 1994*a*). Whether the primary abnormality in these cases is a defect of desmin filaments, or a more general abnormality of protein assembly and degradation remains uncertain.

In severe, X-linked, myotubular myopathy there appears to be arrested maturation of the muscle fibres leading to persistent central nucleation of many fibres in the postnatal period (Sarnat 1990). This is associated with increased expression of desmin, vimentin, and NCAM in the fibres and increased expression of the prenatal myosin isoform (Sawchak *et al.* 1991; Fidzianska and Goebel 1994). In the milder, autosomal dominant form of centronuclear myopathy, the persistent central nucleation of fibres is associated with an abnormal, radial distribution of desmin filaments, and cytoplasmic immunolabelling for dystrophin and spectrin (Fig. 10.10; Sarnat 1990; van der Ven *et al.* 1991; Figarella-Branger *et al.* 1992; Mora *et al.* 1994).

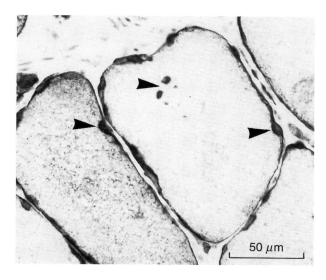

Fig. 10.8 Hereditary distal myopathy with filamentous bodies highlighted by immunolabelling for desmin (arrowheads).

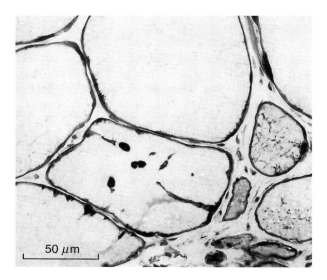

Fig. 10.9 As Fig. 10.8, showing immunolabelling for dystrophin in filamentous bodies.

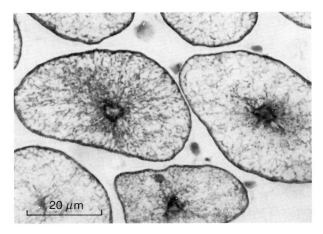

Fig. 10.10 Centronuclear myopathy in a 10–year-old male showing increased desmin immu-nolabelling around the central nuclei and a radial myofibrillar pattern.

10.5.5   Immunohistochemistry of myosin isoforms

In Duchenne dystrophy, myotonic dystrophy, and polymyositis, many fibres co-express fast and slow myosin isoforms or contain molecular hybrids of the two types (Salviati *et al.* 1986). This may be a result of regeneration or of transitions between isoforms occurring in response to changes in neural activity. Myosin immunolabelling becomes more intense and granular during the severe muscular atrophy seen in critically ill patients. Immunohistochemical exam-ination shows that this is accompanied by collapse of the desmin lattice, and electron microscopy reveals a loss of thick filaments, which probably reflects changes in myosin structure during protein catabolism (Wilkinson *et al.* 1995).

10.6   Idiopathic inflammatory myopathies

The diagnosis of dermatomyositis, polymyositis, and inclusion body myositis (IBM) is straightforward when the typical clinical and pathological features occur together (Urbano-Marquez *et al.* 1991). However, when biopsies show little or no abnormality, immunohistochemical studies may provide useful pointers. In both childhood and adult-type dermatomyositis the principal, early, ultrastructural abnormality is capillary endothelial necrosis; muscle fibre degeneration results from ischaemia (Banker 1975; Carpenter *et al.* 1976). Endothelial necrosis can be recognized on light microscopic examination by focal capillary loss on sections labelled for endothelial markers such as *Ulex europaeus*-1 agglutinin, and by clusters of capillaries containing the C5–C9 membrane attack complex, reflecting the immunologically mediated nature of the damage (De Visser *et al.* 1989; Emslie-Smith and Engel 1990; Kissel *et al.* 1991; Estruch *et al.* 1992). In polymyositis and IBM, cell-mediated immune damage by activated T lymphocytes is prominent. Immunocytochemistry at the light and electron microscopic level has shown that the cells that invade non-necrotic fibres are mainly activated cytotoxic/suppressor (T8) lymphocytes, together with a few killer or natural killer cells and macrophages (Rowe *et al.* 1981; Arahata and Engel 1984, 1988; Engel and Arahata 1986; Ringel *et al.* 1987). A similar phenomenon is seen occasionally in Duchenne dystrophy, where it may represent a secondary immune response to damaged muscle components (Arahata and Engel 1988). The lymphocytic infiltrate is accompanied by HLA class 1 antigen and intercellular adhesion molecule-1 (ICAM-1) neo-expression on the surface of invaded and adjacent fibres (Emslie-Smith *et al.* 1989; de Bleeker and Engel 1994). Class 1 antigen expression is a prerequisite for T cell attack but it is not clear that it has a causal role in polymyositis, as it is also seen in dermatomyositis (Emslie-Smith *et al.* 1989), in Duchenne dystrophy (Rowe *et al.* 1983; Appleyard *et al.* 1985), and after exercise-induced damage (Round *et al.* 1987). Nevertheless, the presence of HLA class 1 antigen and of ICAM-1 on the surface of muscle fibres provides a non-specific marker for damage, and is reported to be useful in distinguishing zidovudine myopathy from the inflammatory myopathy associated with AIDS (Gherardi 1994). The role of humoral immunity in polymyositis is also uncertain, although many auto-antibodies occur and antibodies to endothelial cells and to the histidine tRNA synthetase (Jo-1) are both related to interstitial lung disease accompanying polymyositis (Bernstein *et al.* 1984; Cervera *et al.* 1991).

IBM is a relatively benign, slowly progressive myopathy which affects proximal and distal muscles and responds poorly to steroids and immuno-suppressive treatment (Carpenter *et al.* 1978; Calabrese *et al.* 1987). Cytoplasmic, rimmed vacuoles, and 15–18 nm tubulofilamentous cytoplasmic and nuclear inclusions are the pathological hallmarks of the disease, but these are not specific, and there are probably several different myopathies in which inclusion bodies occur, including the Welander form of distal myopathy (Lindberg *et al.* 1991). Paramyxoviruses have now been excluded as possible aetiological agents (Kallajoki *et al.* 1991). The tubulofilamentous inclusions are associated with (but not formed by) ubiquitin and β-amyloid protein (Askanas *et al.* 1992), suggesting that some inclusion body myopathies may be the manifestation of an ageing process akin to Alzheimer's disease.

10.7   Duchenne and Becker dystrophy

10.7.1   Clinical and genetic aspects

Duchenne muscular dystrophy affects 1 in 3000–3500 males, and progresses to the extent that they are usually wheelchair-bound by the age of 12 years and die before the age of 30 years (Gardner-Medwin 1980). Becker dystrophy is a less common and milder form of the disease. The gene responsible for these diseases comprises 70 exons that are distributed over 2500 kilobases of the X chromosome and code for a 427 kD protein, which has been termed dystrophin (Monaco *et al.* 1986; Hoffman *et al.* 1987; Koenig *et al.* 1987; Kunkel and Hoffman 1989). The dystrophin gene is expressed in many tissues including skeletal, smooth, and cardiac muscles; brain; retina; and the myoepithelial cells of sweat and salivary glands (Chelly *et al.* 1988; Miyatake *et al.* 1991).

Deletions of the dystrophin gene occur in 65 per cent of Duchenne and Becker patients and the gene has a high rate of spontaneous mutation (Kunkel and Hoffman 1989). The disease phenotype is not related to the extent of the genetic deletion (Love *et al.* 1989*a*), but it does appear to depend on the C-terminal of dystrophin, which is believed to play an important part in the stability of the molecule. In most patients, disruption of the DNA triplet reading frame results in loss of the C-terminal and a Duchenne phenotype, whereas restoration of the reading frame after gene deletion preserves the C-terminal and leads to a Becker phenotype (Malhotra *et al.* 1988; Monaco *et al.* 1988; Arahata *et al.* 1991; Bulman *et al.* 1991). In Becker dystrophy, the site of the deletion is correlated with the clinical phenotype to some extent: deletions around exons 45/53 cause typical Becker dystrophy, amino terminal deletions are associated with more severe disease, and deletions in the proximal rod domain may lead to myalgia and cramps (Beggs *et al.* 1991). The presence of phenotypic variability within these groups suggests that epigenetic and/or environmental factors play a role in determining clinical progression.

Truncated dystrophin molecules are usually degraded rapidly (Ginjaar *et al.* 1990), although C-terminal deletions are occasionally associated with a truncated dystrophin molecule located in a subsarcolemmal position (Hoffman *et al.* 1991; Helliwell *et al.* 1992*a*; Recan *et al.* 1992). This indicates that the C-terminal is not essential for membrane association, although it is required for proper interaction with dystrophin-associated glycoproteins (Matsumura *et al.* 1993*b, c*; Helliwell *et al.* 1994*c*). Rare cases of N-terminal deletion are associated with preservation of some C-terminal immunoreactivity and a severe Duchenne phenotype (Higuchi *et al.* 1992).

10.7.2 Dystrophin and utrophin immunohistochemistry in the diagnosis of dystrophies

In normal skeletal muscle, dystrophin can be immunolabelled and visualized at the sarcolemma with light or electron microscopy (Arahata *et al.* 1988*a*; Bonilla *et al.* 1988*a*; Shimizu *et al.* 1988; Watkins *et al.* 1988; Zubrzycka-Gaarn *et al.* 1988; Nicholson *et al.* 1989; Ellis *et al.* 1990; Nguyen *et al.* 1990;

Cullen *et al.* 1991). There is concentration of immunolabelling at myotendinous junctions, where dystrophin may be needed for thin filament/membrane interaction (Tidball and Law 1991). In Duchenne dystrophy, dystrophin immunolabelling is absent or reduced (Fig. 10.11), whereas in Becker dystrophy there is weak or discontinuous labelling of the fibre periphery (Arahata *et al.* 1988*a*, 1989; Bonilla *et al.* 1988*a*; Nicholson *et al.* 1989; Vainzoff *et al.* 1990).

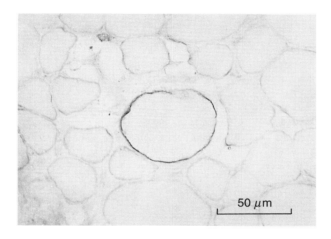

Fig. 10.11 Duchenne dystrophy, showing lack of dystrophin labelling in most fibres. One strongly labelled 'revertant' fibre is present.

Female carriers of Duchenne dystrophy show either a mosaic of positively and negatively labelled fibres (Arahata *et al.* 1988*b*, 1989; Bonilla *et al.* 1988*b*) or, in older patients, variably weak and patchy staining (Clerk *et al.* 1991). Dystrophin-deficient fibres show less dystroglycan labelling and more utrophin labelling than dystrophin-containing fibres (Sewry *et al.* 1994). In Duchenne and Becker dystrophies (and in *mdx* mice), utrophin is expressed extensively at the sarcolemma (Fig. 10.12; Khurana *et al.* 1991; Nguyen *et al.* 1991; Tanaka *et al.* 1991; Voit *et al.* 1991; Helliwell *et al.* 1992*c*). As utrophin has a high degree of structural homology with dystrophin, and can associate with dystrophin-associated glycoproteins in *mdx* muscle, it may be able to compensate in part for dystrophin deficiency (Tanaka *et al.* 1991; Matsumura *et al.* 1992*a*). Dystrophin and utrophin analysis has allowed some female patients with raised creatine kinase levels, non-specific myopathy, or a diagnosis of limb-girdle dystrophy to be reclassified as manifesting carriers of a dystrophin abnormality (Figs 10.13–10.15; Bushby *et al.* 1993; Hoffman *et al.* 1992). The clinician and pathologist must be aware of these presentations so that the correct diagnosis can be made and genetic counselling offered.

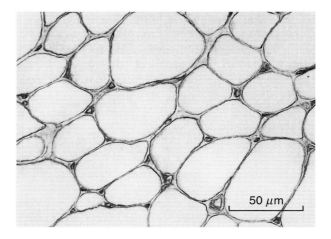

Fig. 10.12 Duchenne dystrophy, showing widespread expression of utrophin on the surface of fibres.

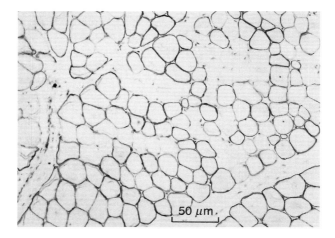

Fig. 10.13 Manifesting female carrier of Duchenne dystrophy, showing groups of fibres that do not label for dystrophin.

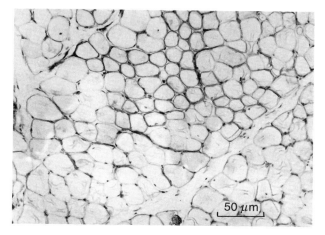

Fig. 10.14   As Fig. 10.13, but labelled for utrophin. Stronger labelling for utrophin occurs in the dystrophin-negative fibres.

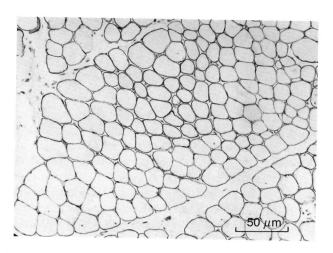

Fig. 10.15   Same section as Figs 10.13 and 10.14. Preservation of the integrity of the sarcolemma in all the fibres is shown by labelling for β-spectrin.

Immunohistochemical labelling of dystrophin localizes native protein in fibres and provides information complementary to Western blotting, which identifies the molecular weight and abundance of denatured protein. Both methods of dystrophin analysis have to be interpreted with care. Biopsy trauma or necrosis can lead to loss of staining, and immunolabelling for β-spectrin should therefore be used as a positive control for membrane integrity (Fig. 10.15). Because deletions may affect any part of the molecule, and because antibodies vary in their binding affinity for dystrophin (Slater and Nicholson 1991), it is necessary to use well characterized antibodies to the N-terminal, rod portion, and C-terminal of the molecule. Unexpected positive labelling in Duchenne dystrophy may occur in cases with a truncated dystrophin molecule (see above) and in those cases with 'revertant' fibres. In many Duchenne biopsies, 0.1–30 per cent of fibres show positive sarcolemmal labelling with both N- and C-terminal antibodies (Fig. 10.11; Shimizu *et al.* 1988; Nicholson *et al.* 1989, 1992; Vainzoff *et al.* 1990). Immunolabelling studies with exon-specific antibodies support the suggestion that these positive fibres probably result from 'somatic reversion', in which a second gene deletion restores the reading frame and allows synthesis of the C-terminal of the molecule (Hoffman *et al.* 1990; Fanin *et al.* 1992; Klein *et al.* 1992; Wallgren-Pettersson *et al.* 1993; Zhao *et al.* 1993; Winnard *et al.* 1995), although different second mutations may operate in different revertant fibres (Thanh *et al.* 1995). In several diseases immunolabelling localizes dystrophin to cytoplasmic granules, which presumably reflect a stage in intracellular protein degradation (Shimizu *et al.* 1988; Nicholson *et al.* 1989; Vainzoff *et al.* 1990). Other abnormalities of dystrophin occur, probably as secondary events, in inflammatory myopathies (Sewry *et al.* 1991). This underlines the need for careful correlation of clinical with histological findings.

10.7.3   Animal models of Duchenne dystrophy

Golden retriever dogs with an X-linked dystrophic process appear to be an excellent model for Duchenne dystrophy (Valentine *et al.* 1986; Kornegay *et al.* 1988). They are more susceptible to exercise-induced damage than normal (McCully *et al.* 1991). They have a mutation of the dystrophin gene (Cooper *et al.* 1988) and show progressive clinical symptoms, cardiomyopathy, and progressive skeletal muscle pathology (Valentine *et al.* 1986, 1988, 1990). Male dogs with manifest disease do not express dystrophin, but female carriers have a mosaic pattern of positive and negative fibres in both skeletal and cardiac muscle (Cooper *et al.* 1990). The mosaicism persists in cardiac muscle, but decreases with age in skeletal muscle, so that all fibres are positive by 24 weeks. A similar phenomenon occurs in the *mdx* mouse (see below), where it is not prevented by irradiation; this suggests that it is the result of accumulation of dystrophin with age rather than the fusion of competent myoblasts (Karpati *et al.* 1990).

Dystrophin-deficient cats suffer from gross muscle hypertrophy which can be lethal when it affects the diaphragm or tongue (Gashen *et al.* 1992).

The C57BL *mdx* strain of mouse was described by Bulfield and colleagues (1984), and subsequently shown to have a point mutation in the dystrophin gene leading to the premature termination of the polypeptide at 27 per cent of its normal length, and loss of expression of immunoreactive dystrophin in blots and in tissue sections (Hoffman *et al.* 1987; Ryder-Cook *et al.* 1988; Sicinski *et al.* 1989). The muscles of *mdx* mice are normal until 14 days old, at which stage hypereosinophilic fibres are seen. Necrosis peaks at 3–5 weeks of age and continues at a low level throughout life. Regeneration is highly effective, although the fibres retain central nuclei, and there is no fibrosis (Bulfield *et al.* 1984; Bridges 1986; Torres and Duchen 1987; Coulton *et al.* 1988). As fibres less than 20 mm diameter (extraocular and denervated fibres) are not susceptible to necrosis, the burst of necrosis at 3–5 weeks may be related to fibre growth beyond a critical point or to increased functional demands (Karpati *et al.* 1988). Intracellular free calcium concentrations in *mdx* muscle are twice those of normal muscle and are associated with increased protein degradation by calcium-dependent thiol proteases (see Chapter 6; Turner *et al.* 1988; MacLennan and Edwards 1990; Dunn and Radda 1991; MacLennan *et al.* 1991). Other degradative processes that may be involved include activation of alkaline (serine) proteases and lysosomal cathepsins (Sawada *et al.* 1986; Sano *et al.* 1988). Despite the histological and biochemical changes, muscle function is well maintained and there is an increase in muscle bulk (Dangain and Vrbová 1984). The *mdx* mouse provides a valuable model in which to study the effects of dystrophin deficiency and the apparently effective compensatory mechanisms (Partridge 1991). In experimental work, however, it is important to recognize that there may be considerable differences in muscle function, and in the extent of necrosis and regeneration, between the different muscles of a given animal, as well as between individual animals of the same and different ages (Dangain and Vrbová 1984; Anderson *et al.* 1988; Geissinger *et al.* 1990; Zacharias and Anderson 1991).

10.7.4 The function of dystrophin and the pathogenesis of Duchenne dystrophy

The discovery of dystrophin was only one step in unravelling the complexity of the molecular processes involved in Duchenne dystrophy. The structure and molecular linkages of dystrophin indicate that abnormalities could compromise the mechanical support of the sarcolemma, links between the cytoskeleton and the extracellular matrix, and the function of other membrane-associated molecules such as ion channels. Since muscle function can be maintained in dystrophin-deficient animals it is also necessary to consider the possible mechanisms by which muscle fibres can compensate for the dystrophin abnormality.

10.7.4.1 *Sarcolemmal support*

If dystrophin contributes to the elastic function of the sub-sarcolemmal filamentous matrix, decreased support for the sarcolemma, leading to increased susceptibility to mechanical damage, would explain the high plasma levels of muscle-derived enzymes in Duchenne dystrophy (Roy and Dubowitz 1970). This is consistent with ultrastructural evidence of discontinuities of the plasma membrane associated with myofibrillar disruption (delta lesions) and increased permeability of the fibres (Mokri and Engel 1975; Schmalbruch 1975; Rowland 1976; Bradley and Fulthorpe 1978; Carpenter and Karpati 1979). However, although some studies in *mdx* mice suggest that dystrophin-deficient muscles are more susceptible to physical stress than normal muscles (Menke and Jockusch 1979; Weller *et al.* 1990; Petroff *et al.* 1993; Pasternak *et al.* 1995), this hypothesis is not supported by other work (Franco and Lansman 1990; Hutter *et al.* 1991; McArdle *et al.* 1991, 1992; Sacco *et al.* 1992). The differences may be attributable in part to the different experimental models used, which place the membrane under different stresses, but the issue is largely unresolved.

10.7.4.2   *Links with the extracellular matrix*

The molecular model for dystrophin suggests close ties with the basement membrane, and hence with the endomysial and perimysial collagen framework. These links are essential for the transmission of the force generated by the myofibrils to the tendons, the passage of signals regulating cell function, and the binding and concentration of soluble cell growth factors (Ruoslahti 1990). The possibility that these links may be disturbed in Duchenne dystrophy is suggested by the abnormal ultrastructure of myotendinous junctions (Law and Tidball 1993), and by the reduced immunolabelling for vinculin and for the dystrophin-associated glycoproteins in Duchenne and *mdx* mouse muscle (Minetti *et al.* 1992*b*; Matsumura *et al.* 1992*a, b*; Ibraghimov-Beskrovnaya *et al.* 1992; Helliwell *et al.* 1994*c*; Matsumura and Campbell 1993; Ohlendieck *et al.* 1993). The mRNA for the glycoproteins is present in normal concentrations, which suggests that in Duchenne dystrophy the glycoproteins are synthesized but are not incorporated stably into the sarcolemma (Ohlendieck *et al.* 1991). It is intriguing that expression of the 43 kD glycoprotein is retained in satellite cells in Duchenne muscle (Fig. 10.16; Helliwell *et al.* 1994*c*). In Becker dystrophy, the expression of dystrophin-associated glycoproteins is reduced to a smaller extent than in Duchenne dystrophy, and correlates closely with the distribution of dystrophin (Matsumura *et al.* 1993*b*; Helliwell *et al.* 1994*c*).

The possible functional effects of disturbing links with the extracellular matrix have been demonstrated in cultures of normal myoblasts: blocking the function of β-integrin allows continued proliferation with no fusion (Menko and Boettiger 1987), whereas TGF-β increases integrin expression and may modulate the binding of cells to the extracellular matrix (Roberts *et al.* 1990).

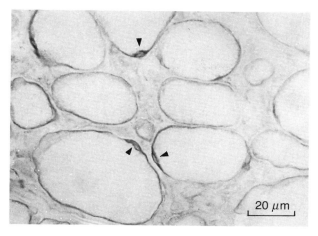

Fig. 10.16 Duchenne dystrophy, showing that immunolabelling for the 43 kD dystrophin-associated glycoprotein is reduced around most of the sarcolemma, although labelling is maintained around presumptive satellite cells (arrowheads).

The basement membrane is synthesized by the muscle fibres, endomysial fibroblasts, and possibly by the endothelial cells (Kuhl *et al.* 1982; Chiquet and Fambrough 1984; Sanderson *et al.* 1986). It is composed of a network of type IV collagen which is linked to laminin, fibronectin, heparan, and chondroitin sulphate proteoglycans, and other proteins such as entactin and nidogen (Abrahamson 1986; Yurchenko 1990). There is little evidence that basement membranes are abnormal in Duchenne dystrophy, although cultured myoblasts show an abnormal surface morphology and reduced synthesis of basement membrane components (Delaporte *et al.* 1990). On tissue sections immunolabelling for type IV collagen and laminin is increased around atrophic fibres owing to reduplicated or redundant basement membrane (Duance *et al.* 1980; Stephens *et al.* 1982). A prominent feature of Duchenne dystrophy is the presence of excess endomysial fibrous tissue (mainly type III collagen) (Duance *et al.* 1980; Stephens *et al.* 1982; Hantai *et al.* 1985; Rampoldi *et al.* 1986). The stimulus for the increased fibrosis is unclear, but could be related to increased production of TGF-β (Massagué *et al.* 1991; Yamazaki *et al.* 1994), or to the presence of mediators derived from mast cells, which are increased in number in dystrophic muscle (Helliwell *et al.* 1990; Gorospe *et al.* 1994). Fibrosis may contribute to the development of irreversible changes in the muscle by adding ischaemic damage to the damage caused by dystrophin deficiency (Partridge 1991).

10.7.4.3 Abnormal ion channels and cell signalling pathways

The location of dystrophin beneath the sarcolemma and its close association with transmembrane proteins are suggestive of a possible role in membrane-bound cell signalling processes (Hardiman 1994; Bork and Sudol 1994, Yang *et al.* 1995). Dystrophin deficiency may affect the structure and function of membrane ion channels leading to a raised intracellular calcium concentration and abnormalities of Na^+/H^+ exchange (Turner *et al.* 1988; Franco and Lansman 1990; Dunn and Radda 1991; Turner *et al.* 1991; Dunn *et al.* 1992, 1993). Muscle fibres in *mdx* mice have a limited capacity for generating ATP; this may also be the result of a lack of stabilization by dystrophin of glycolytic enzyme complexes and mitochondria, which are normally associated with microtubules and the cytoskeleton (Chinet *et al.* 1994).

10.7.4.4   *Reactive/reparative changes*

Since the *mdx* mouse can clearly compensate for dystrophin deficiency, and since dystrophin-negative fibres in animal models and in human carriers do not necessarily undergo necrosis, an understanding of the nature of the reaction of fibres to dystrophin deficiency may be of therapeutic value. Relatively little is known about the compensatory changes, but they may include: stabilization of the sarcolemma by the production of dystrophin homologues (such as utrophin), stress-related proteins (such as HLA-1), and increased amounts of normal subsarcolemmal filamentous proteins, including talin and vinculin (Law *et al.* 1994); maintenance of regenerative activity—possibly by growth factor expression, since basic FGF immunoreactivity is increased in *mdx* muscle (Anderson *et al.* 1991); and a higher rate of protein turnover, prevention of which by nutritional deprivation leads to massive necrosis (Helliwell *et al.* 1992*b*).

In Duchenne dystrophy, the morphology of regeneration and the capacity for regeneration after experimental challenge are similar to those in normal muscle (Walton and Adams 1956; Mussini *et al.* 1987). Regeneration may be less effective after the age of 5–6 years, either because of a reduction in the number of remaining satellite cells or because of premature degeneration of regenerating fibres (Bell and Conen 1967; Mastaglia and Kakulas 1969; Blau *et al.* 1983, 1985).

10.8   New findings in other dystrophies and myopathies

A deficiency of a sarcoglycan (adhalin), has been identified in a severe childhood autosomal recessive muscular dystrophy (SCARMD), described initially in North African patients but since identified in other populations (Matsumura and Campbell 1994). These patients had been diagnosed clinically as Duchenne dystrophy before analysis showed dystrophin to be normal. The a sarcoglycan gene maps to chromosome 17q21 and, while this appears to be the primary abnormality in some families with SCARMD, in other families (including the North African group) the α sarcoglycan abnormality appears to be secondary to genetic abnormalities of the β or γ sarcoglycan genes on chromosomes 13q12 and 4q12, respectively (Campbell 1995; Noguchi *et al.* 1995). Abnormalities of laminin chain expression (reduction of laminin B1 and overexpression of the S chain) are observed in some patients with a sarcoglycan deficiency (Yamada *et al.* 1995), and milder forms of sarcoglycan deficiency are associated with some forms of limb girdle muscular dystrophy (Brown 1996). The BIO 14.6 hamster is a possible animal model for sarcoglycan deficiency (Mizuno *et al.* 1995).

In the Fukuyama-type of congenital muscular dystrophy, dystrophin expression is nearly normal, although there are abnormalities of α and β dystroglycans (Matsumura *et al.* 1993*a*). The disease maps to chromosome 9q31–33, but the primary gene product is not known (Campbell 1995).

The non-Fukuyama-type of congenital muscular dystrophy is a heterogeneous group of severe, early onset, autosomally, recessively inherited muscle diseases. In about 30 per cent of patients there is a specific absence of laminin-α2 (merosin; Tomé *et al.* 1994; Sewry *et al.* 1995; Hoffman 1996), and in other families the disease maps to chromosome 6q2 near the merosin gene (Hillaire *et al.* 1994). A specific defect of laminin-α2 also occurs in the dystrophia muscularis (dy/dy) mouse (Sunada *et al.* 1994).

Identification of the protein molecular abnormality in myotonic dystrophy as overexpression of a membrane-associated protein kinase (Strong *et al.* 1994) raises the possibility that abnormal phosphorylation of other proteins may be important in the pathogenesis of this disease.

The protein product of the gene that is abnormal in Emery-Dreifuss muscular dystrophy has been identified as emerin, a 254–amino acid protein with a transmembrane domain (Bione *et al.* 1994). This will enable another dystrophy to be recognized by immunohistochemical methods.

10.9   Conclusions and future prospects

This chapter has focused on the role of immunohistochemistry in the diagnosis of muscle diseases and in the recognition of pathogenetic mechanisms. A purely biochemical approach to these problems would normally be restricted to the analysis of changes in a relatively large volume of tissue containing a mixture of cell types. This can be extended to the level of single fibres, but only at the cost of many hours of painstaking work on what may then prove to be an unrepresentative fibre population. The use of microscopy, and of immunohistochemical techniques in particular, is therefore a valuable addition to the diagnostic armamentarium of the pathologist. (See Chapter 7 for another way of combining these complementary techniques.)

The pathogenetic mechanisms of Duchenne dystrophy are becoming clearer as the molecular interactions of dystrophin are defined. However, there are still many areas of uncertainty about the precise function or functions of dystrophin, the ways in which dystrophin dysfunction leads to clinical weakness, and the compensatory mechanisms by which *mdx* mice are able to avoid the progressive fibrotic disease that devastates the musculature of patients with Duchenne dystrophy. Immunohistochemical studies have elucidated some aspects of this story and will continue to play a major role, in conjunction with molecular biological techniques, in unravelling the mechanisms of the disease process. As the genetic abnormalities underlying other muscle diseases are defined, so immunohistochemical techniques will be developed to recognize the abnormal proteins and to extend in this way the scope and precision of diagnosis.

In addition to its ability to identify genetic abnormalities in specific proteins, immunohistochemistry can identify specific proteolytic mechanisms in individual fibres. With the development of antibodies to proteases and to specific epitopes on proteins, it should be possible to identify the damaging processes in biopsy and experimental material. This will provide more information of potential therapeutic value and a means of monitoring the response to treatment. Studies of the molecular mechanisms of regeneration and of the role of growth factors should lead to an understanding of the ability of muscle to recover from damage. If this information can be applied therapeutically, it might offer hope for the restoration of muscle function in myopathies.

Acknowledgements

The author wishes to express his appreciation of the support and encouragement of colleagues in the Departments of Medicine and Pathology, University of Liverpool. Grants from the Mersey Regional Health Authority and the Muscular Dystrophy Group of Great Britain and Northern Ireland supported some of the scientific studies described in this chapter and are gratefully acknowledged.

References

Abrahamson, D.R. (1986). Recent studies on the structure and pathology of basement membranes. *Journal of Pathology*, **149**, 257–78.

Akaaboune, M., Villanova, M., Festoff, B.W., Verdiere-Sahuque, M., and Hantai, D. (1994). Apolipoprotein-E expression at neuromuscular junctions in mouse, rat and human skeletal muscle. *FEBS Letters*, **351**, 246–8.

Allbrook, D. (1981). Skeletal muscle regeneration. *Muscle and Nerve*, **4**, 234–45.

Allen, R.E. and Rankin, L.L. (1990). Regulation of satellite cells during skeletal muscle growth and development. *Proceedings of the Society for Experimental Biology and Medicine*, **194**, 81–6.

Anand, R. and Emery, A.E.H. (1980). Calcium stimulated enzyme efflux from human skeletal muscle. *Research Communications in Chemical Pathology and Pharmacology*, **28**, 541–50.

Anderson, J.E., Bressler, B.H., and Ovalle, W.K. (1988). Functional regeneration in the hindlimb skeletal muscle of the *mdx* mouse. *Journal of Muscle Research and Cell Motility*, **9**, 499–515.

Anderson, J.E., Liu, L., and Kardami, E. (1991). Distinctive patterns of basic fibroblast growth factor (bFGF) distribution in degenerating and regenerating areas of dystrophic (*mdx*) striated muscles. *Developmental Biology*, **147**, 96–109.

Appleyard, S.T., Dunn, M.J., Dubowitz, V., Scott, M.L., Pittman, S.J., and Shotton, D.M. (1984). Monoclonal antibodies detect a spectrin-like protein in normal and dystrophic human skeletal muscle. *Proceedings of the National Academy of Science USA*, **81**, 776–80.

Appleyard, S.T., Dunn, M.J., Dubowitz, V., and Rose, M.C. (1985). Increased expression of HLA ABC class 1 antigens by muscle fibres in Duchenne muscular dystrophy, inflammatory myopathies and other neuromuscular disease. *Lancet*, **i**, 361–3.

Arahata, K. and Engel, A.G. (1984). Monoclonal antibody analysis of mononuclear cells in myopathies. I. Quantitation of subsets according to diagnosis and sites of accumulation and demonstration and counts of muscle fibers invaded by T cells. *Annals of Neurology*, **16**, 193–208.

Arahata, K. and Engel, A.G. (1988). Monoclonal antibody analysis of mononuclear cells in myopathies. IV. Cell-mediated cytotoxicity and muscle fiber necrosis. *Annals of Neurology*, **23**, 168–73.

Arahata, K., Ishiura, S., Ishiguro, T., Tsukahara, T., Suhara, Y., Eguchi, C. *et al.* (1988*a*). Immunostaining of skeletal and cardiac muscle surface membrane with antibody against Duchenne muscular dystrophy peptide. *Nature*, **333**, 861–3.

Arahata, K., Ishihara, T., Kamakura, K., Tsukahara, T., Ishiura, S., Baba, C. *et al.* (1988*b*). Mosaic expression of dystrophin in symptomatic carriers of Duchenne's muscular dystrophy. *New England Journal of Medicine*, **320**, 138–42.

Arahata, K., Hoffman, E.P., Kunkel, L.M., Ishiura, S., Tsukahara, T., Ishihara, T. *et al.* (1989). Dystrophin diagnosis, comparison of dystrophin abnormalities by immunofluorescence and immunoblot analyses. *Proceedings of the National Academy of Science USA*, **86**, 7154–8.

Arahata, K., Beggs, A.H., Honda, H., Ito, S., Ishiura, S., Tsukahara, T. *et al.* (1991). Preservation of the C-terminus of dystrophin molecule in the skeletal muscle from Becker muscular dystrophy. *Journal of the Neurological Sciences*, **101**, 148–56.

Askanas, V., Bornemann, A., and Engel, W.K. (1990). Immunocytochemical localization of desmin at human neuromuscular junctions. *Neurology*, **40**, 949–53.

Askanas, V., Engel, W.K., and Alvarez, R.B. (1992). Light and electron microscopic localization of β-amyloid protein in muscle biopsies of patients with inclusion body myositis. *American Journal of Pathology*, **141**, 31–6.

Austin, L. and Burgess, A.W. (1991). Stimulation of myoblast proliferation in culture by leukaemia inhibitory factor and other cytokines. *Journal of the Neurological Sciences*, **101**, 193–7.

Austin, L., Bower, J., Kurek, J., and Vakakis, N. (1992). Effects of leukaemia inhibitory factor and other cytokines on murine and human myoblast proliferation. *Journal of the Neurological Sciences*, **112**, 185–91.

Banker, B.Q. (1975). Dermatomyositis of childhood, ultrastructural alterations of muscle and intramuscular blood vessels. *Journal of Neuropathology and Experimental Neurology*, **33**, 46–75.

Beggs, A.H., Hoffman, E.P., Snyder, J.R., Arahata, K., Specht, L., Shapiro, F. *et al.* (1991). Exploring the molecular basis for variability among patients with Becker muscular dystrophy, dystrophin gene and protein studies. *American Journal of Human Genetics*, **49**, 54–67.

Behr, T., Fischer, P., Muller-Felber, W., Schmidt-Achert, M., and Pongratz, D. (1994).Myofibrillogenesis in primary tissue cultures of adult human skeletal muscle, expression of desmin, titin and nebulin. *Clinical Investigation*, **72**, 150–5.

Bell, C.D. and Conen, P.E. (1967). Change in fiber size in Duchenne muscular dystrophy. *Neurology*, **17**, 902–13.

Bennett, G.S., Fellini, S.A., and Holtzer, H. (1978). Immunofluorescent visualization of 100 Å filaments in different cultured chick embryo cell types. *Differentiation*, **12**, 71–82.

Bernstein, R.M., Morgan, S.H., Chapman, J., Bunn, C.C., Mathews, M.B., and Turner-Warwick, M. (1984). Anti Jo-1 antibody, a marker for myositis with interstitial lung disease. *British Medical Journal*, **289**, 151–2.

Bewick, G.S., Nicholson, L.V.B., Young, C., O'Donnell, E., and Slater, C.R. (1992). Different distributions of dystrophin and related proteins at nerve-muscle junctions. *Molecular Neuroscience*, **3**, 857–60.

Bione, S., Maestrini, E., Rivella, S., Mancini, M., Regis, S., Romeo, G. *et al.* (1994). Identification of a novel X-linked gene responsible for Emery-Dreifuss muscular dystrophy. *Nature Genetics*, **8**, 323–7.

Bischoff, R. (1986). A satellite cell mitogen from crushed adult muscle. *Developmental Biology*, **115**, 140–7.

Bischoff, R. (1990). Control of satellite cell proliferation. *Advances in Experimental Medicine and Biology*, **280**, 147–58.

Blake, D.J., Love, D.R., Tinsley, J., Morris, G.E., Turley, H., Gatter, K. *et al.* (1992). Characterisation of a 4.8 kb transcript from the Duchenne muscular dystrophy locus expressed in Schwannoma cells. *Human Molecular Genetics*, **1**, 103–9.

Blake, D.J., Tinsley, J.M., and Davies, K.E. (1996). Utrophin: a structural and functional comparison to dystrophin. *Brain Pathology*, **6**, 37–47.

Blau, H.M., Webster, C., and Pavlath, G.K. (1983). Defective myoblasts identified in Duchenne muscular dystrophy. *Proceedings of the National Academy of Science USA*, **80**, 4856–60.

Blau, H., Webster, C., Pavlath, G.K., and Chiu, C-P. (1985). Evidence for defective myoblasts in Duchenne muscular dystrophy. *Advances in Experimental Medicine and Biology*, **182**, 85–110.

Bodensteiner, J.B. and Engel, A.G. (1978). Intracellular calcium accumulation in Duchenne dystrophy and other myopathies, a study of 567,000 muscle fibers in 114 biopsies. *Neurology*, **28**, 439–46.

Bolmont, C., Lilienbaum, A., Paulin, D., and Grimaud, J.A. (1990). Expression of desmin gene in skeletal and smooth muscle by *in situ* hybridization using a human desmin gene probe. *Journal of Submicroscopic Cytology and Pathology*, **22**, 117–22.

Bonilla, E., Samitt, C.E., Miranda, A.F., Hays, A.P., Salviati, G., DiMauro, S. *et al.* (1988*a*). Duchenne muscular dystrophy, deficiency of dystrophin at the muscle cell surface. *Cell*, **54**, 447–52.

Bonilla, E., Schmidt, B., Samitt, C.E., Miranda, A.F., Hays, A.P., De Oliveira, A.B.S. *et al.* (1988*b*). Normal and dystrophin-deficient muscle fibers in carriers of the gene for Duchenne muscular dystrophy. *American Journal of Pathology*, **133**, 440–5.

Bork, P. and Sudol, M. (1994). The WW domain, a signalling site in dystrophin? *Trends in Biochemical Sciences*, **19**, 531–3.

Bornemann, A. and Schmalbruch, H. (1991). Desmin and vimentin reactivity in regenerating human and rat muscles. *Clinical Neuropathology*, **10**, 29–30.

Bornemann, A. and Schmalbruch, H. (1993). Anti-vimentin staining in muscle pathology. *Neuropathology and Applied Neurobiology*, **19**, 414–19.

Bornemann, A. and Schmalbruch, H. (1994). Immunocytochemistry of M-cadherin in mature and regenerating rat muscle. *Anatomical Record*, **239**, 119–25.

Bornman, L., Polla, B.S., Lotz, B.P., and Gericke, G.S. (1995). Expression of heat shock/stress proteins in Duchenne muscular dystrophy. *Muscle and Nerve*, **18**, 23–31.

Boxler, K. and Jerusalem, F. (1978). Hyperreactive (hyaline, opaque, dark) muscle fibers in Duchenne dystrophy. *Journal of Neurology*, **219**, 63–72.

Bradley, W.G. (1979). Muscle fibre splitting. In *Muscle regeneration* (ed. A. Mauro), pp. 215–32. Raven Press, New York.

Bradley, W.G. and Fulthorpe, J.J. (1978). Studies of sarcolemmal integrity in myopathic muscle. *Neurology*, **28**, 670–7.

Bridges, L.R. (1986). The association of cardiac muscle necrosis and inflammation with the degenerative and persistent myopathy of mdx mice. *Journal of the Neurological Sciences*, **72**, 147–57.

Brown, R.H. (1996). Dystrophin-associated prtoeins and the muscular dystrophies: a glossary. *Brain Pathology*, **6**, 19–24.

Bulfield, G., Siller, W.G., Wight, P.A.L., and Moore, K.J. (1984). X chromosome-linked muscular dystrophy (mdx) in the mouse. *Proceedings of the National Academy of Science USA*, **81**, 1189–92.

Bulman, D.E., Murphy, E.G., Zubrzycka-Gaarn, E.E., Worton, R.G., and Ray, P.N. (1991). Differentiation of Duchenne and Becker muscular dystrophy phenotypes with amino- and carboxy-terminal antisera specific for dystrophin. *American Journal of Human Genetics*, **48**, 295–304.

Bushby, K.M.D., Goodship, J.A., Nicholson, L.V.B., Johnson, M.A., Haggerty, I.D., and Gardner-Medwin, D. (1993). Variability in clinical, genetic and protein abnormalities in manifesting carriers of Duchenne and Becker muscular dystrophy. *Neuromuscular Disorders*, **3**, 57–64.

Byers, T.B., Kunkel, L.M., and Watkins, S.C. (1991). The subcellular distribution of dystrophin in mouse skeletal, cardiac and smooth muscle. *Journal of Cell Biology*, **115**, 411–21.

Calabrese, L.H., Mitsumoto, H., and Chou, S.M. (1987). Inclusion body myositis presenting as treatment resistant polymyositis. *Arthritis and Rheumatism*, **30**, 397–403.

Campbell, K.P. (1995). Three muscular dystrophies, loss of cytoskeleton-extracellular matrix linkage. *Cell*, **80**, 675–9.

Campbell, K.P. and Kahl, S.D. (1989). Association of dystrophin and an integral membrane glycoprotein. *Nature*, **338**, 259–62.

Carlson, B.M. (1973). The regeneration of skeletal muscle—a review. *American Journal of Anatomy*, **137**, 119–50.

Carpenter, S. and Karpati, G. (1979). Duchenne muscular dystrophy, Plasma membrane loss initiates muscle cell necrosis unless it is repaired. *Brain*, **102**, 147–61.

Carpenter, S. and Karpati, G. (1989). Segmental necrosis and its demarcation in experimental micropuncture injury of skeletal muscle fibers. *Journal of Neuropathology and Experimental Neurology*, **48**, 154–70.

Carpenter, S., Karpati, G., Rothman, S., and Watters, G. (1976). The childhood type of dermatomyositis. *Neurology*, **26**, 952–62.

Carpenter, S., Karpati, G., Heller, I., and Eissen, A. (1978). Inclusion body myositis, a distinct variety of idioathic inflammatory myopathy. *Neurology*, **28**, 8–17.

Cervera, R., Ramirez, G., Fernandez-Sola, J., D'Cruz, D., Casademont, J., Grau, J. *et al.* (1991). Antibodies to endothelial cells in dermatomyositis, association with interstitial lung disease. *British Medical Journal*, **302**, 880–1.

Chelly, J., Kaplan, J.C., Maire, P., Gautron, S., and Kahn, A. (1988). Transcription of the dystrophin gene in human muscle and non-muscle tissues. *Nature*, **333**, 858–60.

Chevron, M.P., Girard, F., Claustres, M., and Demaille, J. (1994). Expression and subcellular localization of dystrophin in skeletal, cardiac and smooth muscles during human development. *Neuromuscular Disorders*, **4**, 419–32.

Chinet, A.E., Even, P.C., and Decrouy, A. (1994). Dystrophin-dependent efficiency of metabolic pathways in mouse skeletal muscles. *Experientia*, **50**, 602–5.

Chiquet, M. and Fambrough, D.M. (1984). Extracellular matrix assembly during muscle development studied with monoclonal antibodies. *Experimental Biology and Medicine*, **9**, 87–92.

Church, J.C.T., Noronha, R.F.X., and Allbrook, D.B. (1966). Satellite cells and skeletal muscle regeneration. *British Journal of Surgery*, **53**, 638–42.

Clerk, A., Rodillo, E., Heckmatt, J.Z., Dubowitz, V., Strong, P.N., and Sewry, C.A. (1991). Characterisation of dystrophin in carriers of Duchenne muscular dystrophy. *Journal of the Neurological Sciences*, **102**, 197–205.

Cooper, B.J., Winand, N.J., Stedman, H., Valentine, B.A., Hoffman, E.P., Kunkel, LM. *et al.* (1988). The homologue of the Duchenne locus is defective in X-linked muscular dystrophy of dogs. *Nature*, **334**, 154–6.

Cooper, B.J., Gallagher, E.A., Smith, C.A., Valentine, B.A., and Winand, N.J. (1990). Mosaic expression of dystrophin in carriers of canine X-linked muscular dystrophy. *Laboratory Investigation*, **62**, 171–8.

Cornelio, F. and Dones, I. (1984). Muscle fiber degeneration and necrosis in muscular dystrophy and other muscle diseases, cytochemical and immunocytochemical data. *Annals of Neurology*, **16**, 694–701.

Coulton, G.R., Morgan, J.E., Partridge, T.A., and Sloper, J.C. (1988). The mdx mouse skeletal muscle myopathy, 1. A histochemical, morphometric and biochemical investigation. *Neuropathology and Applied Neurobiology*, **14**, 53–70.

Cullen, M.J., Walsh, J., Nicholson, L.V.B., Harris, J.B., Zubrzycka-Gaarn, E.E., Ray, P.N. *et al.* (1991). Immunogold labelling of dystrophin in human muscle, using an antibody to the last 17 amino acids of the C-terminus. *Neuromuscular Disorders*, **1**, 113–19.

Cullen, M.J., Fulthorpe, J.J., and Harris, J.B. (1992). The distribution of desmin and titin in normal and dystrophic human muscle. *Acta Neuropathologica*, **83**, 158–69.

Dangain, J. and Vrbová, G. (1984). Muscle development in mdx mutant mice. *Muscle and Nerve*, **7**, 700–4.

Darr, K.C. and Schultz, E. (1987). Exercise-induced satellite cell activation in growing and mature skeletal muscle. *Journal of Applied Physiology*, **63**, 1816–21.

de Bleeker, J.L. and Engel, A.G. (1994). Expression of cell adhesion molecules in inflammatory myopathies and Duchenne dystrophy. *Journal of Neuropathology and Experimental Neurology*, **53**, 369–76.

Delaporte, C., Dautreaux, B., Rouche, A., and Fardeau, M. (1990). Changes in surface morphology and basal lamina of cultured muscle cells from Duchenne muscular dystrophy patients. *Journal of the Neurological Sciences*, **95**, 77–88.

De Visser, M., Emslie-Smith, A.M., and Engel, A.G. (1989). Early ultrastructural alterations in adult dermatomyositis. Capillary abnormalities precede other structural changes in muscle. *Journal of the Neurological Sciences*, **94**, 181–92.

Drachman, D.B., Toyka, K.V., and Myer, E. (1974). Prednisolone in Duchenne muscular dystrophy. *Lancet*, **ii**, 1409–12.

Draeger, A., Weeds, A.G., and Fitzsimmons, R.B. (1987). Primary, secondary and tertiary myotubes in developing skeletal muscle, a new approach to the analysis of human myogenesis. *Journal of the Neurological Sciences*, **81**, 19–43.

Duance, V.C., Stephens, H.R., Dunn, M., Balley, A.J., and Dubowitz, V. (1980). A role for collagen in the pathogenesis of muscular dystrophy? *Nature*, **284**, 470–2.

Duncan, C.J. and Jackson, M.J. (1987). Different mechanisms mediate structural changes and intracellular enzyme efflux following damage to skeletal muscle. *Journal of Cell Sciences*, **87**, 183–8.

Dunn, J.F. and Radda, G.K. (1991). Total ion content of skeletal and cardiac muscle in the mdx mouse dystrophy, Ca^{2+} is elevated at all ages. *Journal of the Neurological Sciences*, **103**, 226–31.

Dunn, J.F., Tracey, I., and Radda, G.K. (1992). A ^{31}P-NMR study of muscle exercise metabolism in mdx mice, evidence for abnormal pH regulation. *Journal of the Neurological Sciences*, **113**, 108–13.

Dunn, J.F., Bannister, N., Kemp, G.J., and Publicover, S.J. (1993). Sodium is elevated in mdx muscles, ionic interactions in dystrophic cells. *Journal of the Neurological Sciences*, **114**, 76–80.

Ecob-Prince, M., Hill, M., and Brown, W. (1989). Immunocytochemical demonstration of myosin heavy chain expression in human muscle. *Journal of the Neurological Sciences*, **91**, 71–8.

Edström, L., Thornell, L-E., and Eriksson, A. (1980). A new type of hereditary distal myopathy with characteristic sarcoplasmic bodies and intermediate (skeletin) filaments. *Journal of the Neurological Sciences*, **47**, 171–90.

Edwards, R.H.T., Isenberg, D.A., Wiles, C.M., Young, A., and Snaith, M.L. (1981). The investigation of inflammatory myopathy. *Journal of the Royal College of Physicians*, **15**, 19–24.

Ellis, J.M., Man, N.T., Morris, G.E., Ginjaar, I.B., Moorman, A.F.M., and Van Ommen, G.J.B. (1990). Specificity of dystrophin analysis improved with monoclonal antibodies. *Lancet*, **336**, 881–2.

Emslie-Smith, A.M. and Engel, A.G. (1990). Microvascular changes in early and advanced dermatomyositis, a quantitative study. *Annals of Neurology*, **27**, 343–56.

Emslie-Smith, A.M., Arahata, K., and Engel, A.G. (1989). Major histocompatability complex class I antigen expression, immunolocalization of interferon subtypes and T cell-mediated cytotoxicity in myopathies. *Human Pathology*, **20**, 224–31.

Engel, A.G. and Arahata, K. (1986). Mononuclear cells in myopathies. Quantitation of functionally distinct subsets, recognition of antigen-specific cell-mediated cytotoxicity in some diseases, and implications for the pathogenesis of the different inflammatory myopathies. *Human Pathology*, **17**, 704–21.

Engel, W.K. (1973). Duchenne muscular dystrophy, a histologically based ischemia hypothesis and comparison with experimental ischemia myopathy. In *The striated muscle* (ed. C.M. Pearson and F.K. Mostofi), pp. 453–72. Williams and Wilkins, Baltimore.

Engel, W.K. (1979). Muscle fibre regeneration in human neuromuscular disease. In *Muscle regeneration* (ed. A. Mauro *et al.*), pp. 285–96. Raven Press. New York.

Engel, W.K. and Cunningham, G.G. (1970). Alkaline phosphatase-positive abnormal muscle fibers of humans. *Journal of Histochemistry and Cytochemistry*, **18**, 55–7.

Ervasti, J.M. and Campbell, K.P. (1991). Membrane organization of the dystrophin-glycoprotein complex. *Cell*, **66**, 1121–31.

Ervasti, J.M., Ohlendieck, K., Kahl, S.D., Gaver, M.G., and Campbell, K,P. (1990). Deficiency of a glycoprotein component of the dystrophin complex in dystrophic muscle. *Nature*, **345**, 315–19.

Estruch, R., Grau, J.M., Fernandez-Sola, J., Casademont, J., Monforte, R., and Urbano-Marquez, A. (1992). Microvascular changes in skeletal muscle in idiopathic inflammatory myopathy. *Human Pathology*, **23**, 888–95.

Ewton, D.Z. and Florini, J.R. (1990). Effects of insulin-like growth factors and transforming growth factor-β on the growth and differentiation of muscle cells in culture. *Proceedings of the Society for Experimental Biology and Medicine*, **194**, 76–80.

Fallon, J.R. and Hall, Z.W. (1994). Building synapses, agrin and dystroglycan stick together. *Trends in Neurosciences*, **17**, 469–73.

Fanin, M., Daniell, G.A., Vitiello, L., Senter, L., and Angelini, C. (1992). Prevalence of dystrophin-positive fibers in 85 Duchenne muscular dystrophy patients. *Neuromuscular Disorders*, **2**, 41–5.

Fidzianska, A. and Goebel, H.H. (1994). Aberrant arrested in maturation neuro-muscular junctions in centronuclear myopathy. *Journal of the Neurological Sciences*, **124**, 83–8.

Figarella-Branger, D., Nedelec, J., Pellisier, J.F., Boucraut, J., Bianco, N., and Rougon, G. (1990). Expression of various isoforms of neural cell adhesive molecules and their highly polysialated counterparts in diseased human muscles. *Journal of the Neurological Sciences*, **98**, 21–36.

Figarella-Branger, D., Calore, E.E., Boucraut, J., Bianco, N., Rougon, G., and Pellisier, J.F. (1992). Expression of cell surface and cytoskeleton developmentally regulated proteins in adult centronuclear myopathies. *Journal of the Neurological Sciences*, **109**, 69–76.

Fishback, D.K. and Fishback, H.R. (1932). Studies of experimental muscle degeneration. 1. Factors in the production of muscle degeneration. *American Journal of Pathology*, **8**, 193–217.

Fitzsimmons, R.B. and Hoh, J.F.Y. (1981). Embryonic and foetal myosins in human skeletal muscle. *Journal of the Neurological Sciences*, **52**, 367–84.

Florini, J.R. and Magri, K.A. (1989). Effects of growth factors on myogenic differentiation. *American Journal of Physiology*, **256**, C701–11.

Forbus, W.D. (1926). Pathologic changes in voluntary muscle. 1, Degeneration and regeneration of the rectus abdominis in pneumonia. *Archives of Pathology*, **2**, 318–39.

Franco, A. and Lansman, J.B. (1990). Calcium entry through stretch-inactivated ion channels in mdx myotubes. *Nature*, **344**, 670–3.

Gardner-Medwin, D. (1980). Clinical features and classification of the muscular dystrophies. *British Medical Bulletin*, **36**, 109–15.

Gashen, F.P., Hoffman, E.P., Gorospe, J.R.M., Uhl, E.W., Senior, D.F., Cardinet, G.H. *et al.* (1992). Dystrophin deficiency causes lethal muscle hypertrophy in cats. *Journal of the Neurological Sciences*, **110**, 149–59.

Gauthier, G.F. and Lowey, S. (1977). Polymorphism of myosin among skeletal muscle fiber types. *Journal of Cell Biology*, **74**, 760–79.

Geissinger, H.D., Prasada-Rao, P.V.V., and McDonald-Taylor, C.K. (1990). "mdx" mouse myopathy, histopathological, morphometric and histochemical observations on young mice. *Journal of Comparative Pathology*, **102**, 249–63.

Geng, Y., Sicinski, P., Gorecki, D., and Barnard, P.J. (1991). Developmental and tissue-specific regulation of mouse dystrophin, the embryonic isoform in muscular dystrophy. *Neuromuscular Disorders*, **1**, 125–33.

Gherardi, R.K. (1994). Skeletal muscle involvement in HIV-infected patients. *Neuropathology and Applied Neurobiology*, **20**, 232–7.

Ginjaar, I.B., Bakker, E., den Dunnen, J.T., Wessels, A., van Paassen, M.M.B., Kloosterman, M.D. *et al.* (1990). Detection of truncated dystrophin in fetal DMD myotubes. *Advances in Experimental Medicine and Biology*, **280**, 17–23.

Goebel, H.H. and Bornemann, A. (1993). Desmin pathology in neuromuscular diseases. *Virchows Archiv B: Cellular Pathology*, **64**, 127–35.

Gorospe, J.R., Tharp, M.D., Hinckley, J., Kornegay, J.N., and Hoffman, E.P. (1994). A role for mast cells in the progression of Duchenne muscular dystrophy? Correlations in dystrophin-deficient humans, dogs and mice. *Journal of the Neurological Sciences*, **122**, 44–56.

Gospodarowicz, D., Weseman, J., Moran, J.S., and Lindstrom, J. (1976). Effect of fibroblast growth factor on the division and fusion of bovine myoblasts. *Journal of Cell Biology*, **70**, 395–405.

Granger, B.L. and Lazarides, E. (1978). The existence of an insoluble Z disc scaffold in chicken skeletal muscle. *Cell*, **15**, 1253–68.

Grounds, M. (1991). Towards understanding skeletal muscle regeneration. *Pathology Research and Practice*, **187**, 1–22.

Gu, M., Wang, W., Song, W.K., Cooper, D.N.W., and Kaufman, S.J. (1994). Selective modulation of the interaction of $\alpha 7 \beta 1$ integrin with fibronectin and laminin by L-14 lectin during skeletal muscle differentiation. *Journal of Cell Science*, **107**, 175–81.

Hantai, D., Labat-Robert, J., Grimaud, J-A., and Fardeau, M. (1985). Fibronectin, laminin, type I, III and IV collagens in Duchenne's muscular dystrophy, congenital muscular dystrophies and congenital myopathies, an immunocytochemical study. *Connective Tissue Research*, **13**, 273–81.

Hardiman, O. (1994). Dystrophin deficiency, altered cell signalling and fibre hypertrophy. *Neuromuscular Disorders*, **4**, 305–15.

Hashimoto, K., Shimizu, T., Nonaka, I., and Mannen, T. (1989). Immunocytochemical analysis of α-actinin of nemaline myopathy after two dimensional electrophoresis. *Journal of the Neurological Sciences*, **93**, 199–209.

Head, S.I., Stephenson, D.G., and Williams, D.A. (1990). Properties of enzymatically isolated skeletal fibres from mice with muscular dystrophy. *Journal of Physiology*, **422**, 351–67.

Helliwell, T.R. (1988). Lectin binding and desmin expression in bupivicaine-induced necrosis and regeneration of rat skeletal muscle. *Journal of Pathology*, **155**, 317–26.

Helliwell, T.R. (1993). The dystrophin-related protein, utrophin, is expressed in atrophic skeletal muscle fibres in spinal muscular atrophy and peripheral neuropathies. *Journal of Pathology*, **169**, 161A.

Helliwell, T.R., Gunhan, O., and Edwards, R.H.T. (1989). Lectin binding and desmin expression during necrosis, regeneration and neurogenic atrophy of human skeletal muscle. *Journal of Pathology*, **159**, 43–51.

Helliwell, T.R., Gunhan, O., and Edwards, R.H.T. (1990). Mast cells in neuromuscular diseases. *Journal of the Neurological Sciences*, **98**, 267–76.

Helliwell, T.R., Ellis, J.M., Montford, R.C., Appleton, R.E., and Morris, G.E. (1992a). A truncated dystrophin lacking the C-terminal domains is localized at the muscle membrane. *American Journal of Human Genetics*, **50**, 508–14.

Helliwell, T.R., MacLennan, P., Jackson, M.J., and Edwards, R.H.T. (1992b). The effect of acute nutritional deprivation on the mdx mouse. *Journal of Pathology*, **167**, 120A.

Helliwell, T.R., Nguyen, thi Man, Morris, G.E., and Davies, K.E. (1992c). The dystrophin-related protein, utrophin, is expressed on the sarcolemma of regenerating human skeletal muscle fibres in dystrophies and inflammatory myopathies. *Neuromuscular Disorders*, **2**, 177–84.

Helliwell, T.R., Green, A.R.T., Green, A., and Edwards, R.H.T. (1994a). Hereditary distal myopathy with granulo-filamentous inclusions containing desmin, dystrophin and vimentin. *Journal of the Neurological Sciences*, **124**, 174–87.

Helliwell, T.R., Jackson, M.J., Phoenix, J., West-Jordan, J., MacLennan, P., and Edwards, R.H.T. (1994b). Immunohistochemical and biochemical indicators of skeletal muscle damage *in vitro*, the stability of control muscle and the effects of damage by dinitrophenol and calcium ionophore. *International Journal of Experimental Pathology*, **75**, 329–43.

Helliwell, T.R., Nguyen, thi Man, and Morris, G.E. (1994c). Expression of the 43kD dystrophin-associated glycoprotein in human neuromuscular disease. *Neuromuscular Disorders*, **4**, 101–13.

Higuchi, I., Fukunaga, H., Usuki, F., Moritoyo, T., and Osame, M. (1992). Phenotypic Duchenne muscular dystrophy with C-terminal domain. *Pediatric Neurology*, **8**, 310–12.

Hillaire, D., Leclerc, A., Faure, S., Topaloglu, H., Chiannilkulchai, N., Guicheney, P. *et al.* (1994). Localization of merosin-negative congenital muscular dystrophy to chromosome 6q2 by homozygosity mapping. *Human Molecular Genetics*, **3**, 1657–61.

Hoffman, E.P. (1996). Clinical and histopathological features of abnormalities of the dystrophin-based membrane cytoskeleton. *Brain Pathology*, **6**, 49–61.

Hoffman, E.P., Arahata, K., Bonilla, E., and Rowland, L.P. (1992). Dystrophino-pathy in isolated cases of myopathy in females. *Neurology*, **42**, 967–75.

Hoffman, E.P., Brown, R.H., and Kunkel, L.M. (1987). Dystrophin, the protein product of the Duchenne dystrophy locus. *Cell*, **51**, 919–28.

Hoffman, E.P., Morgan, J.E., Watkins, S.C., and Partridge, T.A. (1990). Somatic reversion/suppression of the mouse *mdx* phenotype *in vivo*. *Journal of Neuro-logical Sciences*, **99**, 9–25.

Hoffman, E.P., Garcia, C.A., Chamberlain, J.S., Angelini, C., Lupski, J.R., and Fenwick, R. (1991). Is the carboxyl-terminus of dystrophin required for mem-brane association? A novel, severe case of Duchenne muscular dystrophy. *Annals of Neurology*, **30**, 605–10.

Holtzer, H., Bennett, G.S., Tapscott, S.J., Croop, J.M., and Toyama, Y. (1981). Intermediate-size filaments, changes in synthesis and distribution in cells of the myogenic and neurogenic lineages. *Cold Spring Harbour Symposium of Quanti-tative Biology*, **46**, 317–29.

Holtzer, H., Forry-Schaudies, Dlugosz, A., Antin, P., and Dubyak, G. (1985). Interactions between IFs, microtubules, and myofibrils in fibrogenic and myogenic cells. *Annals of the New York Academy of Science*, **455**, 106–25.

Hood, D., van Lente, F., and Estes, M. (1991). Serum enzyme alterations in chronic muscle disease. A biopsy-based diagnostic assessment. *American Journal of Clinical Pathology*, **95**, 402–7.

Horowitz, R., Kempner, E.S., Bisher, M.E., and Podolsky, R.J. (1986). A physio-logical role for titin and nebulin in skeletal muscle. *Nature*, **323**, 160–4.

Hu, D.H., Kimura, S., Kawashima, S., and Maruyama, K. (1989). Calcium-activated neutral protease quickly converts β-connectin to α-connectin in chicken breast muscle myofibrils. *Zoological Science*, **6**, 797–800.

Hutter, O.F., Burton, F.L., and Bovells, D.L. (1991). Mechanical properties of normal and mdx mouse sarcolemma, bearing on function of dystrophin. *Journal of Muscle Research and Cell Motility*, **12**, 585–9.

Ibraghimov-Beskrovnaya, O., Ervasti, J.M., Leveille, C.J., Slaughter, C.A., Sernett, S.W., and Campbell, K.P. (1992). Primary structure of dystrophin-associated glycoproteins linking dystrophin to the extracellular matrix. *Nature*, **355**, 696–702.

Isaacs, E.R., Bradley, W.G., and Henderson, G. (1973). Longitudinal fibre splitting in muscular dystrophy, a serial cinematographic study. *Journal of Neurology, Neurosurgery and Psychiatry*, **36**, 813–19.

Isenberg, G. and Goldmann, W.H. (1992). Actin-membrane coupling, a role for talin. *Journal of Muscle Research and Cell Motility*, **13**, 587–9.

Jackson, M.J., Round, J.M., Newham, D.J., and Edwards, R.H.T. (1987). An examination of some factors influencing creatine kinase in the blood of patients with muscular dystrophy. *Muscle and Nerve*, **10**, 15–21.

Jackson, M.J., Brooke, M.H., Kaiser, K., and Edwards, R.H.T. (1991*a*). Creatine kinase and prostaglandin E_2 release from isolated Duchenne muscle. *Neurology*, **41**, 101–4.

Jackson, M.J., Page, S., and Edwards, R.H.T. (1991*b*). The nature of the proteins lost from isolated rat skeletal muscle during experimental damage. *Clinica Chemica Acta*, **197**, 1–8.

Jennische, E. (1989). Sequential immunohistochemical expression of IGF-1 and the transferrin receptor in regenerating rat muscle. *Acta Endocrinologica*, **121**, 733–8.

Jockusch, B.M., Veldman, H., Griffiths, G., Vanoost, B.A., and Jennekens, F.G.I. (1980). Immunofluorescence microscopy of a myopathy. α-actinin is a major constituent of nemaline rods. *Experimental Cell Research*, **127**, 409–20.

Kallajoki, M., Hyypia, T., Halonen, P., Orvell, C., Rima, B.K., and Kalimo, H. (1991). Inclusion body myositis and paramyxoviruses. *Human Pathology*, **22**, 29–32.

Karpati, G. and Carpenter, S. (1984). Repair of microscopic abnormalities in skeletal muscle fibres. In *Neuromuscular diseases* (ed. G. Serratrice *et al.*), pp. 157–60. Raven Press, New York.

Karpati, G., Carpenter, S., and Eisen, A.A. (1972). Experimental core-like lesions and nemaline rods. *Archives of Neurology*, **27**, 237–51.

Karpati, G., Carpenter, S., and Pena, S. (1981). Tracer and marker techniques in the microscopic study of skeletal muscles. *Methods and Achievements in Experimental Pathology*, **10**, 101–37.

Karpati, G., Carpenter, S., and Prescott, S. (1988). Small-caliber skeletal muscle fibers do not suffer necrosis in mdx mouse dystrophy. *Muscle and Nerve*, **11**, 795–803.

Karpati, G., Zubrzycka-Gaarn, E.E., Carpenter, S., Bulman, D.E., Ray, P.N., and Worton, R.G. (1990). Age-related conversion of dystrophin-negative to -positive fiber segments of skeletal but not cardiac muscle in heterozygote mdx mice. *Journal of Neuropathology and Experimental Neurology*, **49**, 96–105.

Khurana, T.S., Hoffman, E.P., and Kunkel, L.M. (1990). Identification of a chromosome 6–encoded dystrophin-related protein. *Journal of Biological Chemistry*, **265**, 16717–20.

Khurana, T.S., Watkins, S.C., Chafey, P., Chelly, J., Tomé, F.M.S., Fardeau, M. *et al.* (1991). Immunolocalization and developmental expression of dystrophin related protein in skeletal muscle. *Neuromuscular Disorders*, **1**, 185–94.

Kissel, J.T., Halterman, R.K., Rammohan, K.W., and Mendell, J.R. (1991). The relationship of complement-mediated microvasculopathy to the histologic features and clinical duration of disease in dermatomyositis. *Archives of Neurology*, **48**, 26–30.

Klein, C.J., Coovert, D.D., Bulman, D.E., Ray. P.N., Mendell, J.R., and Burghes, A.H.M. (1992). Somatic reversion/suppression in Duchenne muscular dystrophy (DMD), evidence supporting a frame-restoring mechanism in rare dystrophin-positive fibers. *American Journal of Human Genetics*, **50**, 950–9.

Koenig, M., Hoffman, E.P., Bertelson, C.J., Monaco, A.P., Feener, C., and Kunkel, L.M. (1987). Complete cloning of the Duchenne muscular dystrophy (DMD) cDNA and preliminary genomic organization of the DMD gene in normal and affected individuals. *Cell*, **50**, 509–17.

Kornegay, J.N., Tuler, S.M., Miller, D.M., and Levesque, D.C. (1988). Muscular dystrophy in a litter of golden retriever dogs. *Muscle and Nerve*, **11**, 1056–64.

Kuhl, U., Timpl, R., and von der Mark, K. (1982). Synthesis of type IV collagen and laminin in cultures of skeletal muscle cells and their assembly on the surface of myotubes. *Developmental Biology*, **93**, 344–54.

Kunkel, L.M. and Hoffman, E.P. (1989). Duchenne/Becker muscular dystrophy, a short overview of the gene, the protein, and current diagnostics. *British Medical Bulletin*, **45**, 630–43.

Lakonishok, M., Muschler, J., and Horwitz, A.F. (1992). The α5β1 integrin associates with a dystrophin-containing lattice during muscle development. *Developmental Biology*, **152**, 209–20.

Lane, R.J., Emslie-Smith, A., Mosquera, I.E., and Hudgson, P. (1989). Clinical, biochemical and histological responses to treatment in polymyositis, a prospective study. *Journal of the Royal Society of Medicine*, **82**, 333–8.

Law, D.J. and Tidball, J.G. (1993). Dystrophin deficiency is associated with myotendinous junction defects in prenecrotic and fully regenerated skeletal muscle. *American Journal of Pathology*, **142**, 1513–23.

Law, D.J., Allen, D.L., and Tidball, J.G. (1994). Talin, vinculin and DRP (utrophin) concentrations are increased at *mdx* myotendinous junctions following onset of necrosis. *Journal of Cell Science*, **107**, 1477–83.

Lederfein, D., Levy, Z., Augier, N., Mornet, D., Morris, G., Fuchs, O. *et al.* (1992). A 71–kilodalton protein is the major product of the Duchenne muscular dystrophy gene in brain and other nonmuscle tissues. *Proceedings of the National Academy of Science USA*, **89**, 5346–50.

Lindberg, C., Borg, K., Edström, L., Hedström, A., and Oldfors, A. (1991). Inclusion body myositis and Welander distal myopathy, a clinical, neurophysiological and morphological comparison. *Journal of the Neurological Sciences*, **103**, 76–81.

Lipton, B.H. and Schultz, E. (1979). Developmental fate of skeletal muscle satellite cells. *Science*, **205**, 1292–4.

Lotz, H.P. and Engel, A.G. (1987). Are hypercontracted muscle fibers artifacts and do they cause rupture of the plasma membrane? *Neurology*, **37**, 1466–75.

Love, D.R., Forrest, S.M., Smith, T.J., England, S., Flint, T., Davies, K.E. *et al.* (1989*a*). Molecular analysis of Duchenne and Becker muscular dystrophies. *British Medical Bulletin*, **45**, 659–80.

Love, D.R., Hill, D.F., Dickson, G., Spurr, N.K., Byth, B.C., Marsden, R.F. *et al.* (1989*b*). An autosomal transcript in skeletal muscle with homology to dystrophin. *Nature*, **339**, 55–8.

Love, D.R., Morris, G.E., Ellis, J.M., Fairborther, U., Marsden, R.F., Bloomfield, J.F. *et al.* (1991). Tissue distribution of the dystrophin-related gene product and expression in the mdx and dy mouse. *Proceedings of the National Academy of Science USA*, **88**, 3243–7.

Love, D.R., Byth, B.C., Tinsley, J.M., Blake, D.J., and Davies, K.E. (1993). Dystrophin and dystrophin-related proteins, a review of protein and RNA studies. *Neuromuscular Disorders*, **3**, 5–21.

MacDonald, R. and Engel, A.G. (1970). Experimental chloroquine myopathy. *Journal of Neuropathology and Experimental Neurology*, **29**, 479–99.

MacLennan, P.A. and Edwards, R.H.T. (1990). Protein turnover is elevated in muscle of mdx mice *in vitro*. *Biochemical Journal*, **268**, 795–7.

MacLennan, P.A., McArdle, A., and Edwards, R.H.T. (1991). Effects of calcium on protein turnover of incubated muscles from mdx mice. *American Journal of Physiology*, **260**, E594–8.

Malhotra, S.B., Hart, K.A., Klamut, H.J., Thomas, N.S.T., Bodrug, S.E., Burghes, A.H.M. *et al.* (1988). Frame-shift deletions in patients with Duchenne and Becker muscular dystrophy. *Science*, **242**, 755–9.

Maltin, C.A. and Delday, M.I. (1992). Satellite cells in innervated and denervated muscles treated with clenbuterol. *Muscle and Nerve*, **15**, 919–25.

Massagué, J., Cheifetz, S., Endo, T., and Nadal-Ginard, B. (1986). Type transforming growth factor is an inhibitor of myogenic differentiation. *Proceedings of the National Academy of Science USA*, **83**, 8206–10.

Massagué, J., Heino, J., and Laiho, M. (1991). Mechanisms of TGF-β action. *CIBA Symposium*, **157**, 51–65.

Mastaglia, F.L. and Kakulas, B.A. (1969). Regeneration in Duchenne muscular dystrophy, a histological and histochemical study. *Brain*, **92**, 809–18.

Matsumura, K. and Campbell, K.P. (1993). Deficiency of dystrophin-associated proteins, a common mechanism leading to muscle cell necrosis in severe childhood muscular dystrophies. *Neuromuscular Disorders*, **3**, 109–18.

Matsumura, K. and Campbell, K.P. (1994). Dystrophin-glycoprotein complex, its role in the molecular pathogenesis of muscular dystrophies. *Muscle and Nerve*, **17**, 2–15.

Matsumura, K., Ervasti, J.M., Ohlendieck, K., Kahl, S.D., and Campbell, K.P. (1992a). Association of dystrophin-related protein with dystrophin-associated glycoproteins in mdx mouse muscle. *Nature*, **360**, 588–91.

Matsumura, K., Tomé, F.M.S., Collin, H., Azibi, K., Chaouch, M., Kaplan, J.-C. et al. (1992b). Deficiency of the 50K dystrophin-associated glycoprotein in severe childhood autosomal recessive muscular dystrophy. *Nature*, **359**, 320–2.

Matsumura, K., Nonaka, I., and Campbell, K.P. (1993a). Abnormal expression of dystrophin-associated proteins in Fukuyama-type congenital muscular dystrophy. *Lancet*, **341**, 521–2.

Matsumura, K., Nonaka, I., Tomé, F.M.S., Arahata, K., Collin, H., Leturcq, F. et al. (1993b). Mild deficiency of dystrophin-associated proteins in Becker muscular dystrophy patients having in-frame deletions in the rod domain of dystrophin. *American Journal of Human Genetics*, **53**, 409–16.

Matsumura, K., Tomé, F.M.S., Ionasescu, V., Ervasti, J.M., Anderson, R.D., Romero, N.B. et al. (1993c). Deficiency of dystrophin-associated proteins in Duchenne muscular dystrophy patients lacking COOH-terminal domains of dystrophin. *Journal of Clinical Investigation*, **92**, 866–71.

Mauro, A. (1961). Satellite cell of skeletal muscle fibers. *Journal of Biophysiology, Biochemistry and Cytology*, **9**, 493–5.

McArdle, A., Edwards, R.H.T., and Jackson, M.J. (1991). Effects of contractile activity on muscle damage in the dystrophin-deficient mdx mouse. *Clinical Science*, **80**, 367–71.

McArdle, A., Edwards, R.H.T., and Jackson, M.J. (1992). Accumulation of calcium by normal and dystrophin-deficient mouse muscle during contractile activity *in vitro*. *Clinical Science*, **82**, 455–9.

McCully, K., Giger, U., Argov, Z., Valentine, B., Cooper, B., Chance, B. et al. (1991). Canine X-linked muscular dystrophy studied with *in vivo* phosphorus magnetic resonance spectroscopy. *Muscle and Nerve*, **14**, 1091–8.

Menke, A. and Jockusch, H. (1991). Decreased osmotic stability of dystrophin-less muscle cells from the mdx mouse. *Nature*, **349**, 69–71.

Menko, A.S. and Boettiger, D. (1987). Occupation of the extracellular matrix receptor, integrin, is a control point for myogenic differentiation. *Cell*, **51**, 51–7.

Miner, J.H. and Wold, B.J. (1991). c-myc inhibition of MyoD and myogenin-initiated myogenic differentiation. *Molecular Cell Biology*, **11**, 2842–51.

Minetti, C., Beltrame, F., Marcenaro, G., and Bonilla, E. (1992a). Dystrophin at the plasma membrane of human muscle fibers shows a costameric localization. *Neuromuscular Disorders*, **2**, 99–109.

Minetti, C., Tanjii, K., and Bonilla, E. (1992*b*). Immunologic study of vinculin in Duchenne muscular dystrophy. *Neurology*, **42**, 1751–4.

Miyatake, M., Miike, T., Zhao, J.-E., Yoshioka, K., Uchino, M., and Usuku, G. (1991). Dystrophin, localization and presumed function. *Muscle and Nerve*, **14**, 113–19.

Mizuno, Y., Noguchi, S., Yamamoto, H., Yoshida, M., Nonaka, I., Hirai, S. *et al.* (1995). Sarcoglycan complex is selectively lost in dystrophic hamster muscle. *American Journal of Pathology*, **146**, 530–6.

Mokri, B. and Engel, A.G. (1975). Duchenne dystrophy, electron microscopic findings pointing to a basic or early abnormality in the plasma membrane of the muscle fiber. *Neurology*, **25**, 1111–20.

Monaco, A.P., Neve, R.L., Colletti-Feener, C., Bertelson, C.J., Kurnit, D.M., and Kunkel, L.M. (1986). Isolation of candidate cDNAs for portions of the Duchenne muscular dystrophy gene. *Nature*, **323**, 646–50.

Monaco, A.P., Bertelson, C.J., Liechti-Gallati, S., Moser, H., and Kunkel, L.M. (1988). An explanation for the phenotypic differences between patients bearing partial deletions of the DMD locus. *Genomics*, **2**, 90–5.

Moore, S.E., Hurko, O., and Walsh, F.S.(1984). Immunocytochemical analysis of fibre type differentiation in developing skeletal muscle. *Journal of the Neurological Sciences*, **7**, 137–49.

Mora, M., Morandi, L., Merlini, L., Vita, G., Badadello, A., Barresi, R. *et al.* (1994). Fetus-like dystrophin expression and other cytoskeletal protein abnormalities in centronuclear myopathies. *Muscle and Nerve*, **17**, 1176–84.

Moss, F.P. and Leblond, C.P. (1971). Satellite cells as the source of nuclei in muscles of growing rats. *Anatomical Record*, **170**, 421–36.

Mundegar, R.R., von Oertzen, J., and Zierz, S. (1995). Increased laminin A expression in regenerating myofibers in neuromuscular disorders. *Muscle and Nerve*, **18**, 992–9.

Mussini, I., Favaro, G., and Carraro, U.C. (1987). Maturation, dystrophic changes and the continuous production of fibers in skeletal muscle regenerating in the absence of nerve. *Journal of Neuropathology and Experimental Neurology*, **46**, 315–31.

Nelson, W.J. and Lazarides, E. (1983). Expression of the beta subunit of spectrin in non-erythroid cells. *Proceedings of the National Academy of Science USA*, **80**, 363–7.

Nelson, W.J. and Traub, P. (1981). Properties of a Ca^{2+}-activated protease specific for the intermediate-sized filament protein vimentin in Ehrlich-ascites-tumour cells. *European Journal of Biochemistry*, **116**, 51–7.

Nguyen, thi Man, Cartwright, A.J., Morris, G.E., Love, D.R., Bloomfield, J.F., and Davies, K.E. (1990). Monoclonal antibodies against defined regions of the muscular dystrophy protein, dystrophin. *FEBS Letters*, **262**, 237–40.

Nguyen, thi Man, Ellis, J.M., Love, D.R., Davies, K.E., Gatter, K.C., Dickson, G. *et al.* (1991). Localization of the DMDL gene-encoded dystrophin-related protein using a panel of nineteen monoclonal antibodies. *Journal of Cell Biology*, **115**, 1695–700.

Nicholson, L.V.B., Davison, K., Johnson, M.A., Slater, C.R., Young, C., Bhattacharya, S. *et al.* (1989). Dystrophin in human muscle. 2. Immunoreactivity in patients with Xp21 muscular dystrophy. *Journal of the Neurological Sciences*, **94**, 137–46.

Nicholson, L.V.B., Johnson, M.A., Davison, K., O'Donnell, E., Falkous, G., Barron, M. *et al.* (1992). Dystrophin or a 'related protein' in Duchenne muscular dystrophy? *Acta Neurologica Scandinavica*, **86**, 8–14.

Noguchi, S., McNally E.M., Othmane K.B., Hagiwara Y., Mizuno, Y., Yoshida, M. *et al.* (1995). Mutations in the dystrophin-associated protein sarcoglycan in chromosome 13 muscular dystrophy. *Science*, **270**, 819–22.

Nonaka, I. and Sugita, H. (1980). Muscle pathology of Duchenne dystrophy—with particular reference to "opaque" fibers. *Advances in Neurological Science*, **24**, 718–28.

Ohlendieck, K., Ervasti, J.M., Matsumura, K., Kahl, S.D., Leveille, C.J., and Campbell, K.P. (1991). Dystrophin-related protein is localized to neuromuscular junctions of adult skeletal muscle. *Neuron*, **7**, 499–508.

Ohlendieck, K., Matsumura, K., Ionasescu, V.V., Towbin, J.A., Bosch, E.P., Weinstein, S.L. *et al.* (1993). Duchenne muscular dystrophy, deficiency of dystrophin-associated proteins in the sarcolemma. *Neurology*, **43**, 795–800.

Olive, M. and Ferrer, I. (1994). Parvalbumin immunohistochemistry in denervated skeletal muscle. *Neuropathology and Applied Neurobiology*, **20**, 495–500.

Orimo, S., Arahata, K., Hiyamuta, E., and Sugita, H. (1990). Early breakdown of the membrane associated cytoskeletal proteins, dystrophin and spectrin in bupivicaine induced acute muscle fiber necrosis. *Journal of the Neurological Sciences*, **98S**, 229.

Osborn, M. and Goebel, H.H. (1983). The cytoplasmic bodies in a congenital myopathy can be stained with antibodies to desmin, the muscle-specific intermediate filament protein. *Acta Neuropathologica*, **62**, 149–52.

Partridge, T. (1991). Animal models of muscular dystrophy—what can they teach us? *Neuropathology and Applied Neurobiology*, **17**, 353–63.

Pasternak, C., Wong, S., and Elson, E.L. (1995). Mechanical function of dystrophin in muscle cells. *Journal of Cell Biology*, **128**, 355–61.

Petroff, B.J., Sgrager, J.B., Stedman, H.H., Kelly, A.M., and Sweeney, H.L. (1993). Dystrophin protects the sarcolemma from stresses developed during muscle contraction. *Proceedings of the National Academy of Science USA*, **90**, 3710–14.

Pierobon-Bormioli, S., Sartore, S., Libera, L.D., Vitadello, M., and Schiaffino, S. (1981). "Fast" isomyosins and fiber types in mammalian skeletal muscle. *Journal of Histochemistry and Cytochemistry*, **29**, 1179–88.

Pierobon-Bormioli, S., Sartore, S., Vitadello, M., and Schiaffino, S. (1980). "Slow" myosins in vertebrate skeletal muscle. *Journal of Cell Biology*, **85**, 672–81.

Pons, F., Leger, J.O.C., Chevallay, M., Tomé, F.M.S., Fardeau, M., and Leger, J.J. (1986). Immunocytochemical analysis of myosin heavy chains in human fetal skeletal muscles. *Journal of the Neurological Sciences*, **76**, 151–63.

Pons, F., Augier, N., Leger, J.O.C., Robert, A., Tomé, F.M.S., Fardeau, M. *et al.* (1991). A homologue of dystrophin is expressed at the neuromuscular junctions of normal individuals and DMD patients, and of normal and mdx mice. Immunological evidence. *FEBS Letters*, **282**, 161–5.

Porte, A., Stoeckel, M.-E., Sacrez, A., and Batzenschlager, A. (1980). Unusual familial cardiomyopathy with storage of intermediate filaments in the cardiac muscular cells. *Virchows Archiv (A) Pathology and Anatomy*, **386**, 43–58.

11

Damage in functional grafts of skeletal muscle

Stanley Salmons

11.1 Introduction

There is nothing new about transposing skeletal muscles for cosmetic repair; such procedures are part of the routine armoury of the reconstructive plastic surgeon. If, however, the motor nerve to the muscle is carefully preserved during the procedure, and stimulated electrically after transposition, the potential exists for a *functional* muscle graft. This presents a new range of therapeutic possibilities, of which two will be considered here.

11.1.1 Cardiac assistance from skeletal muscle

Over the last 17 years a significant new surgical approach to the treatment of end-stage heart failure has emerged, based on the redeployment of skeletal muscle. The muscle of choice is latissimus dorsi, which can be transferred into the chest with the main neurovascular bundle intact. In principle, there a variety of ways in which the muscle could then be arranged to assist the heart, and a number of these are the subject of ongoing research (reviewed by Salmons and Jarvis 1992). In the configurations that have been used clinically, the sheet of muscle is wrapped around existing structures: the ventricles of the heart (*cardiomyoplasty*) or the ascending or descending aorta (*aortomyoplasty*). Experience with aortomyoplasty is too limited, in terms of both the number of patients (about 15 worldwide at the time of writing) and the duration of follow-up, to judge its value at this stage. Cardiomyoplasty, on the other hand, has been carried out in over 600 patients worldwide since it was first introduced in 1985 (Carpentier and Chachques 1985), and there is now a substantial body of follow-up data (Grandjean *et al.* 1991; Carpentier *et al.* 1993; Magovern *et al.* 1993; El Oakley and Jarvis 1994; Hagège *et al.* 1995). Some 80 per cent of patients experience symptomatic benefit after the operation, although there is still disagreement about the way in which this benefit is achieved (Hooper and Salmons 1993; El Oakley and Jarvis 1994; Hagège *et al.* 1995; Kass *et al.* 1995; Schreuder *et al.* 1995).

The power available for assisting the heart can be harnessed more effectively if the latissimus dorsi muscle is configured as a separate auxiliary pump, or skeletal muscle ventricle. This application, still undergoing experimental evaluation, illustrates dramatically the potential for functional muscle grafts. In work from Dr L.W. Stephenson's laboratory in Detroit, for example, skeletal muscle ventricles have pumped in circulation in dogs for over 2 years (Mocek *et al.* 1992; Thomas *et al.* 1996) and some are currently functioning as diastolic counterpulsators after more than 3 years (L.W. Stephenson, personal communication).

11.1.2   Neoanal sphincters from skeletal muscle

Chronic faecal incontinence can be the result of congenital anomalies, pudendal neuropathy, trauma to the anal sphincter, or resection of the sphincter during surgery for colorectal malignancy. Patients with this condition may now be offered an alternative to the standard medical and surgical repertoire of treatments. In the new procedure, the patient's gracilis muscle is detached at its distal end and arranged to encircle the anal canal, forming a new anal sphincter (Baeten *et al.* 1988; Williams *et al.* 1989; Cavina *et al.* 1990). A similar 'gracilis sling' or *graciloplasty* operation was introduced some years ago (reviewed by Christiansen *et al.* 1990), but the addition of electrical stimulation enables the graft to function more effectively.

11.1.3   The problem of graft viability

It is clear that the ultimate success of such applications depends on the ability of the grafted muscle to maintain its contractile function in the long term. We should therefore be concerned about the potential challenge to the integrity of the graft posed by transposition, reconfiguration, and electrical stimulation. Indeed, animal studies show these concerns to be justified (Anderson *et al.* 1988; Radermecker *et al.* 1991; Cheng *et al.* 1993; Lucas *et al.* 1993; El Oakley *et al.* 1995). Damage is seen when such grafted muscles are examined after many months, the typical pattern being an increase in intra- and interfascicular fat and fibrous connective tissue, with a corresponding loss of muscle tissue. In some cases the remaining muscle fibres appear normal: in others, they are smaller and more rounded than usual, and may have one or more large vacuoles in the cytoplasm, lending them a moth-eaten appearance. Studies in man are more limited but there is some evidence, from biochemical markers (Moreira *et al.* 1993) and nuclear magnetic resonance imaging (Kalil-Filho *et al.* 1994), that similar changes may take place in the human latissimus dorsi muscle used as a cardiomyoplasty wrap. It is by no means certain that such damage is sustained in every case; indeed, instances have been reported in which samples taken from the muscle *post mortem* had a normal histological appearance (Chachques, personal communication). However, one suspects that damage may be present in a proportion of cases, particularly perhaps in the 20 per cent or so of cardiomyoplasty patients that do not appear to derive any benefit from the procedure. Several factors could contribute to this damage and although their effects would be expected to be combinatorial it will be helpful to consider them separately.

11.2   Effect of stimulation

11.2.1   Differences between natural and artificial activation of muscle

The fibres that make up a muscle are not controlled individually but as motor units: groups of fibres innervated by a single motor neurone. In muscles that contain many hundreds of motor units, such as latissimus dorsi, the most important way in which force is graduated is by the recruitment of more or less motor units. This is not a random process: low-level contractions are brought about by small motor units and, as the force increases, motor units of progressively larger size become involved (Henneman *et al.* 1965a, b, 1974; Stuart and Enoka 1983). The small motor units, which are active for much of the time, are composed of slow, type 1 fibres, whose properties are appropriate for sustaining tension without fatigue for long periods. The larger motor units, which are activated less frequently, are composed of fast, type 2 fibres, whose properties are suited to generating more powerful contractions but on an intermittent basis. This basic organization of the motor system minimizes the possibility of fatigue, because activities such as posture that pose a continuous demand for energy are directed to motor units that are well adapted for sustained use.

The orderly recruitment of motor units is determined through a hierarchy of firing thresholds in the corresponding spinal motor neurones, and it is therefore lost (or even inverted) when attempts are made to exploit the functional capacity of the muscle by electrical stimulation of the motor nerve. This may create a mismatch between the level of work demanded by a particular stimulation regime and the energy that a fibre can supply on a continuous basis through oxidative metabolism.

Other factors exacerbate this mismatch. Under physiological conditions, muscle contraction is brought about by the asynchronous activity of many motor units. Electrical stimulation, on the other hand, activates these motor units synchronously, and it is necessary to use a higher frequency to produce a smooth contraction. Since each impulse is associated with the release and active re-accumulation of calcium ions, this places a greater energy burden on the muscle. Furthermore, under physiological conditions the larger, less frequently used motor units are protected by a natural strategy for minimizing fatigue that is inherent in the pattern of motor neurone firing. Force develops rapidly in response to two or three impulses delivered at a high frequency, after which it is maintained by impulses delivered at a much lower frequency. Such sophistication is not a feature of existing commercially available neuromuscular stimulators, although recent work may encourage efforts in this direction (Kwende *et al.* 1995).

In the light of these differences between physiological and artificial stimulation of muscle, it is little wonder that early attempts to pioneer cardiac assistance from skeletal muscle were thwarted by muscle fatigue, which developed within minutes or hours. New interest in this approach was prompted by the discovery that continuous activation over a period of weeks induces adaptive changes in the muscle, which transform its properties and increase dramatically its resistance to fatigue (Salmons and Sréter 1976). In the long term, therefore, a muscle can be activated through the unphysiological agency of an electrical stimulator because it becomes conditioned to the new demands, even when those demands are equivalent to continuously sustained cardiac work (Acker *et al.* 1987). The question we are chiefly concerned with in this chapter is: what happens in the short term?

11.2.2 Early metabolic events in the response to stimulation

Chronic stimulation at a constant frequency of 10 Hz is similar to the sustained neural impulse activity associated with postural function (Vrbová 1963; Hennig and Lømo 1985). When such a pattern is imposed on muscle fibres that are normally recruited only intermittently, the demand for energy exceeds the supply. It would be anticipated that under these conditions high-energy phosphate would decline rapidly, resulting in severe cellular damage. It has been confirmed that ATP and phosphocreatine do indeed fall to very low levels in the first 5 min of stimulation (Hood and Parent 1991; Mayne *et al.* 1991; Green *et al.* 1992; Salmons *et al.* 1996). However, the incidence of damage in muscles that have been stimulated *in situ* is quite low. In rabbit muscles that were examined 8–10 days after the onset of stimulation, at which stage the histological manifestations of damage are maximal (Maier *et al.* 1986), more than 85 per cent of the fibres had a completely normal appearance (see Chapter 5), and this rose to 100 per cent at 3 weeks as a result of regenerative processes. In rats there is no evidence of damage at any stage, even when the frequency is as high as 20 Hz (Jarvis *et al.* 1996). The reasons for this remarkable resistance to damage are not certain, but there is some evidence that a muscle can decouple itself from a potentially damaging level of use by blocking contraction, both at the level of neuromuscular transmission and at the level of excitation-contraction coupling (Hood and Parent 1991; Mayne *et al.* 1991; Green *et al.* 1992). Intracellular ATP and phosphocreatine are then substantially restored, regaining over 70 per cent of their control levels after 1 hour, despite ongoing stimulation. Palpation suggests that the block is not complete, and that the stimulated muscle is contracting normally again after a few days.

Changes in the levels of other key metabolites provide an insight into the metabolic adjustments that take place during the first minutes and hours of stimulation (Henriksson *et al.* 1988; Hood and Parent 1991; Green *et al.* 1991, 1992; Salmons *et al.* 1996). Although the levels of ATP and phospho-creatine begin to recover after the first 15 min of stimulation, glycogen continues to fall to very low levels. This fall is probably due to rapid glycogenolysis, for it is associated with a rise in glucose under conditions in which hexokinase, and therefore entry of exogenous glucose, should be inhibited. After 1–2 hours glycogen begins to increase, regaining control levels by 4 days. During this phase of the response it would appear that large amounts of glucose are entering the cells—possibly facilitated by a marked increase in hexokinase activity (Weber and Pette 1990)—and that this is being utilized both for synthesizing glycogen and for generating energy by aerobic glycolysis (Green *et al.* 1991). These early metabolic changes contain a number of puzzling features in regard to the regulation of the associated enzymes. For example, glycogen resynthesis takes place despite low levels of the glycogen synthetase stimulator, glucose-6–phosphate, and high levels of the glycolysis stimulator, glucose-1,6–bisphosphate (Salmons *et al.* 1996). Although these phenomena may be illuminated by further study, there seems to be little doubt that the handling of exogenous glucose plays a pivotal role in the early stages of the response of a muscle to stimulation.

11.2.3 Compounding factors

For muscles that are stimulated *in situ* there is variation in the level of damage between species, between individual animals, between different anatomical muscles, and between different patterns of stimulation (see Chapter 5). However, the extent of damage is altogether greater if the muscle is mobilized prior to stimulation, particularly if it is then reattached at less than its normal resting length (El Oakley *et al.* 1995). Stimulation then adds substantially to the level of damage that would have been sustained through mobilization alone. We will return to this aspect later (see §11.4).

11.2.4 The case for minimizing damage

As noted above, and in more detail in §§1.7 and 10.4.2, muscle has considerable potential for regeneration. There are, however, good reasons for wanting to minimize any damage incurred by stimulation. Firstly, damage may lead to a temporary decline in function, which could be important in the context of cardiac assistance, especially in the immediate postoperative period. Secondly, damage liberates proteins such as myoglobin and creatine phosphokinase into the circulation, which may have undesirable systemic effects. Thirdly, some foci of damage could exceed the local capacity of the muscle for self-repair, leaving a residue of connective tissue.

The conditioning protocol that has been widely adopted for clinical cardiomyoplasty (Grandjean *et al.* 1991) represents an attempt to limit stimulation-induced damage by gradually escalating the pattern over a period of 8 weeks. While such measures are intuitively reasonable, their postulated benefits have never been put to a formal test. This said, there is a precedent for such a practice from exercise studies: it has been shown in both animals and man that the susceptibility of a muscle to exercise-induced damage can be reduced by a prior bout of exercise (Clarkson and Tremblay 1988; Ebbeling and Clarkson 1989; Balnave and Thompson 1993; Bosman *et al.* 1993; see also §1.6.2). We therefore conducted a experiment to ascertain: (1) whether such an effect could be produced with stimulation; and (2) whether this could be achieved without the actual induction of damage by the prior stimulation (Jones *et al.* 1996). Previous work (reviewed in Chapter 5) had established that stimulation with a continuous train of impulses at 2.5 Hz did not, of itself, cause significant damage. We therefore prestimulated rabbit ankle extensor muscles with this pattern and then challenged the muscles with a regime (10 Hz, 24 h/day, for 9 days) that usually produces some histological damage (Table 5.1). These experiments confirmed that prestimulation did indeed render the muscles less susceptible to damage by the more intensive regime. Damage was still significant for the protocol that began with 2 days of prestimulation, but as the period of prestimulation was extended to 7 and 14 days, damage was reduced to insignificant levels (Fig. 11.1).

The question arises as to whether prestimulation initiates events along the same pathway as ischaemic preconditioning,[1] the effect of single or repeated short ischaemic episodes on the ability of myocardium to survive prolonged ischaemia and reperfusion without damage (Murry *et al.* 1986; Yellon *et al.* 1993; Marber 1994). Preconditioning can also be brought about by exposing animals to hyperthermia, which appears to be associated with the expression of heat shock proteins (HSPs) and other stress-related proteins (Das *et al.* 1993). A similar phenomenon may occur in skeletal muscle, since exercise is

---

[1] Editor's footnote. In this book, the word 'preconditioning' is used in several contexts and with different meanings. In §9.3.2 it refers to repetitive stretches that lead to mechanical stabilization. The preconditioning discussed here refers to possible ways of protecting muscle (including cardiac muscle) from ischaemic damage by exposing it to prior heat shock or ischaemic stress. Later in this chapter 'preconditioning' is used to describe the induction of adaptive changes in a skeletal muscle prior to mobilizing it as a graft. Despite these various uses, no serious confusion should result: in all contexts the word carries the sense of modifying the properties of the muscle in preparation for a forthcoming change in conditions.

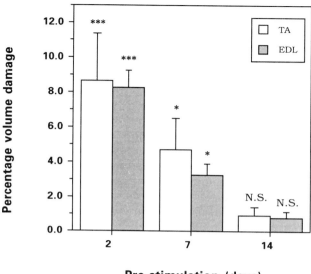

Pre-stimulation (days)

Fig. 11.1 The volume percentage of damage induced in tibialis anterior (TA) and extensor digitorum longus (EDL) muscles of the rabbit by chronic indirect stimulation for 9 days at 10 Hz after prestimulation for 2, 7, and 14 days at 2.5 Hz ($n = 5$ for each group). Significant differences from controls ($n = 14$) are denoted: ***, $p < 0.001$; *, $p < 0.05$. N.S., Not significantly different to control muscle. (Reproduced, with permission, from Jones *et al.* 1996.)

reported to produce a marked rise in HSP 70 and other stress proteins in limb muscles; this could be due to the sublethal elevation in temperature or to oxidative stress (Polla *et al.* 1991; Salo *et al.* 1991; see also §6.2.3). However, this expression of stress proteins occurs within hours of the triggering event (Salo *et al.* 1991). In the present experiments even 2 days of prestimulation was without obvious benefit. It seems that the duration of prestimulation needs to be between 7 and 14 days for a protective effect to be observed. This time course points away from a role of HSPs and other stress-related proteins in the protective influence of prestimulation demonstrated here, but would be consistent with a partial vascular and metabolic adaptation that enabled the muscle to withstand more successfully the subsequent increase in demand. Heat shock or ischaemic preconditioning could, of course, form the basis of other approaches to protection (Ianuzzo *et al.* 1996).

11.3 Effect of partial loss of vascular supply

The blood supply of the latissimus dorsi muscle comes from two main sources. One of these, the thoracodorsal artery, is not disturbed when the muscle is mobilized as a pedicled flap. The other, consisting of branches of intercostal arteries that perforate the chest wall and enter the muscle more distally, has to be ligated and divided in order to mobilize the muscle. Division of these so-called collateral vessels has been shown to result in damage, particularly in the distal portion of the muscle, in both sheep (El Oakley *et al.* 1995) and goats (Ianuzzo *et al.* 1996).

The apparent vascular insufficiency that results from the loss of the collateral blood supply would be expected to exacerbate the mismatch between the continuous level of energy demanded by chronic stimulation and the energy that it is capable of sustaining through oxidative metabolism. This is not to say that the cellular events that precipitate damage are related entirely to metabolic overload, but any component of damage that is of ischaemic origin is likely to be more important in the mobilized muscle. To overcome this problem, Mannion *et al.* (1986, 1989) introduced a vascular delay of 3 weeks before commencing stimulation. The clinical cardiomyoplasty protocol already mentioned (Grandjean *et al.* 1991) also incorporates a vascular delay, in this case of 2 weeks' duration, although this was intended partly to allow adhesions to develop between the muscle flap and the epicardium. Even after such a vascular delay, however, the additive damaging effects of stimulation and division of collateral vessels could be demonstrated in both of the studies referred to above (El Oakley *et al.* 1995; Ianuzzo *et al.* 1996).

Chronic stimulation of a muscle *in situ* induces a dramatic increase in capillary density within 1 week (Hudlicka *et al.* 1982) and there is clearly potential for using this stimulation-induced angiogenesis to promote vascularization of a mobilized muscle. The timing is critical, however, as introducing stimulation too soon, or at too demanding a level, could contribute further to damage rather than avert it.

An alternative strategy would be to commence stimulation some time before mobilizing the muscle. A rational foundation for this approach is provided by experiments in which a period of chronic stimulation was followed by a period of quiescence to trace the time course with which the muscle regained its original properties (Brown *et al.* 1989). The dense capillarization that had developed in the chronically stimulated muscles took many weeks to revert, and would be expected to outlast a short period of postoperative recovery. This raises again the question as to whether it may be advantageous to perform the grafting procedure clinically with a preconditioned flap (Bitto *et al.* 1986). Should this be the case, it would be prudent to look at ways of achieving prior stimulation in a minimally invasive way, rather than to subject a patient to the trauma of an additional full operative procedure.

In the meantime, surgical groups in Paris, France and Brescia, Italy have reported promising results with a pre-operative programme of specific exercise for the latissimus dorsi muscle in patients awaiting cardiomyoplasty. At the very least, exercise might reverse some of the deterioration that has been observed in the skeletal muscles of patients with long-standing heart failure (see Clark and Coats 1995; Scelsi and Scelsi 1995). Moreover, patients could well derive psychological benefit from a not-too-demanding daily routine that contributes to the success of their operation.

Recently we have drawn attention to an important feature of the blood supply to the latissimus dorsi muscle (Craven *et al.* 1994; Degens *et al.* 1996*b*). Using a resin-casting technique in the rat and the rabbit, we were able to demonstrate the proximal and distal arterial trees, representing the branches of the thoracodorsal and perforating arteries respectively, but we made an additional observation: injection via either route appeared to fill the entire vascular network, even when the other route was ligated. Since the resin had to be injected under pressure, it was possible that artefactual channels had been created. We therefore studied the blood supply to the latissimus dorsi muscle under more physiological conditions with the use of fluorescent microspheres. The experiments were conducted on anaesthetized sheep. Fluorescent microspheres were introduced directly into the left ventricle via a carotid catheter and flow was calibrated by withdrawing blood at a fixed rate from a catheter in one femoral artery. Measurements were made during stimulation-induced hyperaemia in the left latissimus dorsi muscle to ensure conditions of maximal blood flow. The right muscle remained at rest, and served as a control for the haemodynamic stability of the animal. We made three determinations of flow, using microspheres labelled with different dyes: normal flow; flow with the thoracodorsal artery clamped; and flow with the thoracodorsal artery patent but with the collateral vessels ligated and divided. Figure 11.2 illustrates the mean of three such experiments. The results show that the territory supplied by the thoracodorsal artery extends over the whole muscle, but diminishes from proximal to distal, whereas the territory supplied by the penetrating branches of the intercostal arteries extends over the whole muscle, but diminishes from distal to proximal.

These observations are consistent with earlier reports of anastomotic channels linking the two vascular territories in this muscle (Tatjanchenko and Sherstennikov 1978; Mathes and Nahai 1981; Tobin *et al.* 1981; Taylor and Palmer 1987; Radermecker *et al.* 1992). The significance of these findings is that the distal portion of the mobilized muscle could, in principle, continue to be perfused via the thoracodorsal artery—despite the loss of the perforating arteries—via an *existing* vascular network.

Future work will need to be focused on ways of enhancing the flow to the distal part of the graft from the thoracodorsal artery. One interesting approach, which appears to have a beneficial effect on graft survival, depends on the injection of growth factors, such as fibroblast growth factor and transforming growth factor β, to promote angiogenesis in the muscle (Mannion *et al.* 1996).

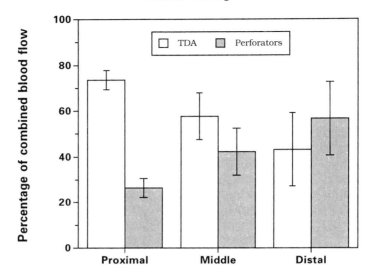

Portion of latissimus dorsi muscle

Fig. 11.2 Intramuscular blood flow distribution in unconditioned sheep latissimus dorsi muscle, measured by injection of fluorescent microspheres under hyperaemic conditions. Flow from the thoracodorsal artery (TDA) alone, and flow from perforating branches of intercostal arteries alone are shown as a percentage of the combined flow, for proximal, middle, and distal portions of the muscle. Mean of three experiments. Note that the distal portion of the muscle receives a substantial contribution from the thoracodorsal artery.

11.4   Effect of loss of tension

The tension developed by a muscle when it is activated depends on the distance between its attachments, and is maximal when the muscle is at its physiological length, that is, with the limb in the position of habitual use. It would therefore be desirable to reconfigure the graft in a manner that preserved its physiological length. In practice this is difficult to achieve. Although the relaxed muscle tends to be thought of as slack, the tension associated with the resting length is actually quite substantial. If a cardiomyoplasty wrap were made at this tension, it would impede refilling of the ventricles during diastole. Alternative ways of using skeletal muscle for cardiac assistance (reviewed by Salmons and Jarvis 1992) do not entirely overcome the problem. Consider, for example, a graft configured as a skeletal muscle ventricle (Acker *et al.* 1987; Mocek *et al.* 1992; Thomas *et al.* 1996). The wall of such a ventricle has a pronounced curvature, which means that fibres within its thickness are stretched to different extents so that, although some fibres could reach physiological levels of resting tension during the filling part of the cycle, these conditions could not be achieved throughout the wall. In another proposed configuration the muscle would be attached in a linear manner to some form of mechanical actuator or generator. In this case the difficulty of creating a strong and reliable junction between muscle or tendon and a non-biological surface would make it unlikely that physiological resting tension could be established, at least in the first instance; even tendon-to-tendon anastomoses take many months to acquire full strength.

For these reasons, it is almost inevitable that part or all of the grafted muscle will be at a resting tension below the physiological level. One consequence of this is that the muscle operates less efficiently, at a point below the optimum on the tension-length curve. A more important consideration is that a lack of stretch depresses protein synthesis, and an unloaded muscle is therefore prone to atrophy (Goldberg *et al.* 1974). Some of the histopathological features seen in grafted muscles are not unlike those seen in muscles whose tendons have been cut (McMinn and Vrbová 1964; Baker 1985; see also §10.5.1), and the causal factors in each case may include a lack of adequate tension. Such a muscle may be less able to respond to both the short-term and the long-term demands of electrical stimulation.

This line of reasoning is well illustrated by experiments alluded to earlier. In that study (El Oakley *et al.* 1995), the effects of different stages of the cardiomyoplasty procedure were studied in a sheep model. The latissimus dorsi muscles of one group of animals were mobilized and reattached at 80 per cent of their original length, at which setting resting tension is minimal. When this intervention was combined with stimulation, 60 per cent of the muscle was found to have been replaced by fibrofatty tissue after 3 months. One may conclude that it is important for the graft to be placed under adequate tension, insofar as this is compatible with other requirements.

This said, other phenomena may serve to balance out some of the deleterious effects of transferring a muscle graft at reduced tension. Muscles that are stimulated chronically *in situ* undergo a 3–5–fold increase in protein synthesis (Salmons 1987). Intermittent mechanical stretch is known to have a stimulatory effect on muscle growth (Vandenburgh *et al.* 1989; Goldspink *et al.* 1995). Such conditions may well be experienced by a muscle that is wrapped around the heart or configured as a skeletal muscle ventricle. Furthermore, in the long term the resting length of surviving fibres in the grafted muscle may self-optimize by remodelling, since it is known that sarcomeres are added to the ends of fibres that are too stretched and removed from the ends of fibres that are too slack (Tabary *et al.* 1972). All these influences may offset those of inappropriate (but probably not grossly inadequate) resting tension.

11.5   Effect of intramuscular pressure

When a muscle contracts against a substantial load, the intramuscular pressure increases. If the pressure rises above that in the intramuscular vessels, the latter will be compressed, so that nutrient blood flow is reduced or even blocked for the duration of the contraction (Petrofsky and Hendershot 1984). In normal use, this effect goes unnoticed: a forceful contraction is followed by relaxation, during which there is a compensatory rush of blood, and the mean blood flow is unchanged (Hussain and Magder 1991). When, however, a muscle is activated incessantly, as in cardiac assistance, there is so little time for recovery that mean blood flow could be depressed, with an accompanying risk of ischaemic damage.

In a recent study (Degens *et al.* 1996*a*), force, intramuscular pressure, and blood flow were measured in rabbit tibialis anterior muscles during repetitive isometric contractions at duty cycles that were varied between 10 and 40 per cent. Intramuscular pressure was measured with a slit-catheter; mean blood flow during stimulation was measured by the method of coloured dye-extraction microspheres; the duty cycle was varied by changing the repetition rate, maintaining the train duration at 100 ms. Stimulation always produced a large increase in blood flow over resting values, and the effect was not attenuated as the duty cycle was increased up to 40 per cent. Mean blood flow under these conditions was not related to integrated pressure, nor to integrated force, and did not differ significantly between the superficial and deep layers of the muscle.

These results seem to suggest that mean blood flow is not compromised by elevated intramuscular pressure for duty cycles up to 30–40 per cent. Even so, the blood flow may not be sufficient to meet the energy demand created at the higher duty cycles. This has been shown in another recent study (Doorn *et al.* 1995). Sheep latissimus dorsi muscles were preconditioned *in situ* with 2–Hz stimulation continuously for 8 weeks, effecting a type transformation that is known to include an increase in capillary density and a switch to a predominantly oxidative metabolism. In a terminal procedure, each animal was reoperated to expose the neurovascular pedicle; an ultrasonic flow probe was placed around the thoracodorsal artery and the vein was cannulated to obtain serial measurements of serum lactate concentration. The muscle was then challenged with bursts of impulses at 32 Hz, with a duration corresponding to either 21 per cent or 30–35 per cent of the electrocardiographic R-R interval. When stimulation was delivered on alternate cardiac cycles (1:2 mode) there was a 2–3–fold functional hyperaemia and a fall in venous lactate concentration. The results were the same whether stimulation was timed to coincide with diastole or systole. However, when stimulation was delivered in 1:1 mode, the venous lactate concentration rose and the functional hyperaemia was followed by a reactive hyperaemia. These experiments show that 1:1 burst stimulation has the potential for causing hypoxic damage.

These studies may provide a somewhat conservative assessment of the problem, for two reasons. Firstly, they were conducted in muscles whose collateral blood supply had not been disturbed. Secondly, the muscles were allowed to contract *in situ*, under which conditions the latissimus dorsi muscle develops a force that is essentially parallel to the direction of its constituent fibres. When the same muscle is configured as a wrap, there is a substantial component of force perpendicular to the fibre axes; it is this component that is responsible for the squeezing effect of the wrap. As a result, fibres in the inner layers of the wrap are compressed by fibres in the outer layers. This effect is more pronounced when the muscle is wound in more than one layer with a small radius of curvature—conditions found, for example, in an aortomyoplasty wrap or in the wall of a skeletal muscle ventricle. In these situations, intramuscular pressure changes associated with contraction would be expected to be larger than those associated with the muscle *in situ*, and the effects on blood flow and tissue oxygenation may be more marked.

Clinical experience appears to bear out these conclusions. Mention has already been made (§11.1.3) of a cardiomyoplasty series in which long-term damage in the latissimus dorsi muscle wrap was observed by magnetic resonance imaging (Kalil-Filho *et al.* 1994); this series was conducted with a 1:1 assist ratio. In most other centres, the practice is to restrict the assist ratio to 1:2. In Brescia there is a preference for 1:4, and the protocol that appeared to produce the best results in the Russian cardiomyoplasty program did not exceed 1:4 (Chekanov 1996). For neoanal sphincters these arguments may be more important still, for here the demand is for low-level, but more continuous, force production from the muscle graft.

11.6 Summary: minimizing damage in functional skeletal muscle grafts

In this chapter we have considered several factors which, alone or in combination, could produce damage in a functional skeletal muscle graft. It is worth emphasizing that muscle damage is not a *necessary* component of the conditioning process. Moreover, some morphological abnormalities that would be included in any histological definition of damage—such as central nuclei, hypereosinophilic fibres, and even moth-eaten fibres—may have little bearing on the function of a graft. It would seem sensible to distinguish these from forms of damage—including invasion by mononuclear cells and replacement by fat or connective tissue—that have functional consequences.

As we are not in a position to assess the relative importance of the various factors, it seems prudent to take account of all of them. Ideally the graft should be placed under stretch, although surgical considerations may force a compromise on this condition. Introducing stimulation at the earliest possible stage has the potential advantage of promoting angiogenesis and of offsetting the catabolic state that results from reduced tension. This objective must be balanced against the potentially injurious effects of stimulating prematurely a muscle that may be inadequately vascularized and inadequately stretched. The problem may be alleviated by optimizing the pattern of stimulation, introducing it in an appropriately staged manner, and, where feasible, preconditioning the graft before it is lifted. Although the early metabolic adjustments to stimulation are not fully understood, exogenous glucose appears to play a key role, and patients undergoing muscle conditioning may therefore benefit from a high-carbohydrate diet.

Even after the conditioning process has been completed it will be necessary to set an upper limit to the duty cycle of stimulation for a particular graft configuration, both to avoid prolonged elevations of intramuscular pressure that could reduce the mean blood flow to the tissue, and to ensure that the demand for oxygen and nutrients does not exceed the supply. Without such precautions the long-term viability of the graft could be compromised.

The therapeutic application of functional skeletal muscle grafts is an exciting new technology. By understanding and addressing the issues discussed in this chapter we will be able to put it to the best possible use.

Acknowledgements

This chapter owes much to exchanges of ideas between the author and his research colleagues. Particular thanks go to Dr J.C. Jarvis for many stimulating discussions, and to Dr H. Degens, Miss H. Sutherland, Mr S.J. Gilroy, Mr M.M.N. Kwende, Miss A.J. Craven, and Mrs M. Hastings. Some of the material was drawn from research collaborations with Drs A. Shortland and R.A. Black (Department of Clinical Engineering); Mr T.L. Hooper and Ms C.A.M. van Doorn (Wythenshawe Hospital, Manchester); Dr D.J. Hitchings, Dr S.R.W. Grainger, Mr I. Taylor, and Mr T. Gunning (Medical Electronics Research Group, University of Staffordshire); Dr L.W. Stephenson (Harper Hospitals, Detroit); and Dr O.H. Lowry and his colleagues at the Washington University School of Medicine, St Louis, Missouri. The support of The British Heart Foundation, The Beit Memorial Foundation, The Wellcome Trust, the Engineering and Physical Sciences Research Council, and the European Community is gratefully acknowledged.

References

Acker, M.A., Hammond, R.L., Mannion, J.D., Salmons, S., and Stephenson, L.W. (1987). Skeletal muscle as the potential power source for a cardiovascular pump: assessment in vivo. *Science,* **236**, 324–7.

Anderson, W.A., Andersen, J.S., Acker, M.A., Hammond, R.L., Chin, A.J., Douglas, P.S. *et al.* (1988). Skeletal muscle grafts applied to the heart: a word of caution. *Circulation,* **78** (Suppl. III), 180–90.

Baeten, C., Spaans, F., and Fluks, A. (1988). An implanted neuromuscular stimulator for faecal continence following previously implanted gracilis muscle: report of a case. *Diseases of the Colon and Rectum,* **31**, 134–7.

Baker, J.H. (1985). The development of central cores in both fibre types in tenotomized muscle. *Muscle and Nerve,* **8**, 115–19.

Balnave, C.D. and Thompson, M.W. (1993). Effect of training on eccentric exercise-induced muscle damage. *Journal of Applied Physiology,* **75**, 1545–51.

Bitto, T., Mannion, J.D., Hammond, R., Cox, J., Yamashita, J., Duckett, S.W. *et al.* (1986). Preparation of fatigue-resistant diaphragmatic muscle grafts for myo-cardial replacement. In *Progress in artificial organs* (ed. Y. Nose, C. Kjellstrand, and P. Ivanovich), pp. 441–6. ISAO Press, Cleveland.

Bosman, P.J., Balemans, W.A.F., Amelink, G.J., and Bär, P.R. (1993). A single training session affects exercise-induced muscle damage in the rat. In *Neuro-muscular fatigue* (ed.A.J. Sargeant and D. Kernell), pp. 74–5. North-Holland, Amsterdam.

Brown, J.M.C., Henriksson, J., and Salmons, S. (1989). Restoration of fast muscle characteristics following cessation of chronic stimulation: physiological, histo-chemical and metabolic changes during slow-to-fast transformation. *Proceedings of the Royal Society of London, Series B,* **235**, 321–46.

Carpentier, A. and Chachques, J.-C. (1985). Myocardial substitution with a stimu-lated skeletal muscle: first successful clinical case. *The Lancet,* **i**, 1267.

Carpentier, A., Chachques, J.-C., Acar, C., Relland, J., Mihaileanu, S., Bensasson, D. *et al.* (1993). Dynamic cardiomyoplasty at seven years. *Journal of Thoracic and Cardiovascular Surgery,* **106**, 42–54.

Cavina, E., Seccia, M., Evangelista, G., Chiarugi, M., Buccianti, P., Tortora, A. *et al.* (1990). Perineal colostomy and electrostimulated gracilis "neosphincter" after abdominoperineal resection of the colon and anorectum: a surgical experience and follow-up study in 47 cases. *International Journal of Color-ectal Disease,* **5**, 6–11.

Chekanov, V.S. (1996). The 1:4 stimulation regimen is the most effective for cardiomyoplasty (clinical evidence). *First Meeting of the International Func-tional Electrical Stimulation Society,* Cleveland, May 1996.

Cheng, W., Michele, J.J., Spinale, F.G., Sink, J.D., and Santamore, W.P. (1993). Effects of cardiomyoplasty on biventricular function in canine chronic heart failure. *Annals of Thoracic Surgery,* **55**, 893–901.

Christiansen, J., Sorensen, M., and Rasmussen, O.O. (1990). Gracilis muscle transposition for faecal incontinence. *British Journal of Surgery,* **77**, 1039–40.

Clark, A.L. and Coats, A.J.S. (1995). Changes in lower limb skeletal muscle and mechanisms of fatigue in chronic heart failure. *Basic and Applied Myology,* **5**, 349–58.

Clarkson, P.M. and Tremblay, I. (1988). Exercise-induced muscle damage, repair, and adaptation in humans. *Journal of Applied Physiology,* **65**, 1–6.

Craven, A.J., Jarvis, J.C., and Salmons, S. (1994). Vascularisation of the latissimus dorsi muscle for cardiac assist. *Journal of Anatomy,* **185**, 706–7.

Das, D.K., Engelman, R.M., and Kimura, Y. (1993). Molecular adaptation of cellular defenses following preconditioning of the heart by repeated ischaemia. *Cardiovascular Research*, **27**, 578–84.

Degens, H., Salmons, S., and Jarvis, J.C. (1996*a*). Mean blood flow and intramuscular pressure during cyclic activity in the rabbit tibialis anterior muscle. In *Proceedings of the Concerted Action "HEART": Skeletal Muscle Assist Working Group Meeting, Florence, Italy* (ed. S. Salmons), pp. 9–13. Commission of the European Communities, Brussels.

Degens, H., Tang, A.T.M, Jarvis, J.C., Hastings, M., and Salmons, S. (1996*b*). The vascular territories in the sheep latissimus dorsi muscle. *First Meeting of the International Functional Electrical Stimulation Society*, Cleveland, May 1996.

Doorn, C.A.M. van, Bhabra, M.S., Hopkinson, D., Greenhalgh, D., and Hooper, T.L. (1995). Effects of latissimus dorsi muscle stimulation patterns on muscle blood flow. In *Proceedings of the Concerted Action "HEART": Skeletal Muscle Assist Working Group Meeting, Maastricht, The Netherlands* (ed. S. Salmons), pp. 15–16. Commission of the European Communities, Brussels.

Ebbeling, C.B. and Clarkson, P.M. (1989). Exercise-induced muscle damage and adaptation. *Sports Medicine*, **7**, 207–34.

El Oakley, R.M. and Jarvis, J.C. (1994). Cardiomyoplasty: a critical review of experimental and clinical results. *Circulation*, **90**, 2085–90.

El Oakley, R.M., Jarvis, J.C., Barman, D., Greenhalgh, D.L., Currie, J., Downham, D.Y. *et al.* (1995). Factors affecting the integrity of latissimus dorsi muscle grafts: implications for cardiac assistance from skeletal muscle. *Journal of Heart and Lung Transplantation*, **14**, 359–65.

Goldberg, A.L., Jablecki, C., and Li, J.B. (1974). Effects of use and disuse on amino acid transport and protein turnover in muscle. *Annals of the New York Academy of Sciences*, **228**, 190–201.

Goldspink, D.F., Cox, V.M., Smith, S.K., Eaves, L.A., Osbaldeston, N.J., Lee, D.M. *et al.* (1995). Muscle growth in response to mechanical stimuli. *American Journal of Physiology*, **268**, E288–97.

Grandjean, P.A., Lori Austin, R.N., Chan, S., Terpestra, B., and Bourgeois, I.M. (1991). Dynamic cardiomyoplasty: clinical follow up results. *Journal of Cardiac Surgery*, **6**, 80–8.

Green, H.J., Cadefau, J., and Pette, D. (1991). Altered glucose 1,6–bisphosphate and fructose 2,6–bisphosphate levels in low-frequency stimulated rabbit fast-twitch muscle. *FEBS Letters*, **282**, 107–9.

Green, H.J., Düsterhöft, S., Dux, L., and Pette, D. (1992). Metabolite patterns related to exhaustion, recovery and transformation of chronically stimulated rabbit fast-twitch muscle. *Pflügers Archiv European Journal of Physiology*, **420**, 359–66.

Hagège, A.A., Desnos, M., Fernandez, F., Besse, B., Mirochnik, N., Castaldo, M. *et al.* (1995). Clinical study of the effects of latissimus dorsi muscle flap stimulation after cardiomyoplasty. *Circulation*, **92**, II-210–15.

Henneman, E., Somjen, G., and Carpenter, D.O. (1965*a*). Functional significance of cell size in spinal motoneurones. *Journal of Neurophysiology*, **28**, 560–80.

Henneman, E., Somjen, G., and Carpenter, D.O. (1965*b*). Excitability and inhibitability of motoneurones of different sizes. *Journal of Neurophysiology*, **28**, 599–620.

Henneman, E., Clamann, H.P., Gillies, J.D., and Skinner, R.D. (1974). Rank order of motoneurons within a pool: law of combination. *Journal of Neurophysiology*, **37**, 1338–49.

Hennig, R. and Lømo, T. (1985). Firing patterns of motor units in normal rats. *Nature, London*, **314**, 164–6.

Henriksson, J., Salmons, S., Chi, M.M.-Y., Hintz, C.S., and Lowry, O.H. (1988). Chronic stimulation of mammalian muscle: changes in metabolite concentrations in individual fibers. *American Journal of Physiology*, **255**, C543–51.

Hood, D.A. and Parent, G. (1991). Metabolic and contractile responses of rat fast-twitch muscle to 10–Hz stimulation. *American Journal of Physiology*, **260**, C832–40.

Hooper, T.L. and Salmons, S. (1993). Skeletal muscle assistance in heart failure. *Cardiovascular Research*, **27**, 1404–6.

Hudlicka, O., Dodd, L., Renkin, E., and Gray, S.D. (1982). Early changes in fiber profile and capillary density in long-term stimulated muscles. *American Journal of Physiology*, **243**, H528–35.

Hussain, S.N.A. and Magder, S. (1991). Diaphragmatic intramuscular pressure in relation to tension, shortening, and blood flow. *Journal of Applied Physiology*, **71**,159–67.

Ianuzzo, C.D., Ianuzzo, S.E., Carson, N., Feild, M., Locke, M., Gu, J. *et al.* (1996). Cardiomyoplasty: degeneration of the assisting skeletal muscle. *Journal of Applied Physiology*, **80**, 1205–13.

Jarvis, J.C., Mokrusch, T., Kwende, M.M.N., Gilroy, S., Sutherland, H., and Salmons, S. (1996). Fast-to-slow transformation in stimulated rat muscle. *Muscle and Nerve*, **19**, 1469–75.

Jones, J., Emmanuel, J., Sutherland, H., Jackson, M.J., Jarvis, J.C., and Salmons, S. (1996). Stimulation-induced skeletal muscle damage: cytoprotective effect of prestimulation. *Basic and Applied Myology*, **7**, in press.

Kalil-Filho, R., Bocchi, E., Weiss, R.G., Rosemberg, L., Bacal, F., Moreira, L.F.P. *et al.* (1994). Magnetic resonance imaging evaluation of chronic changes in latissimus dorsi cardiomyoplasty. *Circulation*, **90**, II-102–6.

Kass, D.A., Baughman, K.L., Pak, P.H., Cho, P.W., Levin, H.R., Gardner, T.J. *et al.* (1995). Reverse modeling from cardiomyoplasty in human heart failure. External constraint versus active assist. *Circulation*, **91**, 2314–18.

Kwende, M.M.N., Jarvis, J.C., and Salmons, S. (1995). The input-output relationships of skeletal muscle. *Proceedings of the Royal Society of London, Series B*, **261**, 193–201.

Lucas, C.M.H.B., van der Veen, F.H., Cheriex, E.C., Lorusso, R., Havenith, M., Penn, O.C.K.M. *et al.* (1993). Long-term follow-up (12 to 35 weeks) after dynamic cardiomyoplasty. *Journal of the American College of Cardiologists*, **22**, 758–67.

Magovern, J.A., Magovern, G.J., Maher, T.D. Jr, Benckart, D.H., Park, S.B., Christlieb, I.Y. *et al.* (1993). Operation for congestive heart failure: transplantation, coronary artery bypass, and cardiomyoplasty. *Annals of Thoracic Surgery*, **56**, 418–25.

Maier, A., Gambke, B., and Pette, D. (1986). Degeneration-regeneration as a mechanism contributing to the fast to slow conversion of chronically stimulated fast-twitch rabbit muscle. *Cell and Tissue Research*, **244**, 635–43.

Mannion, J.D., Hammond, R.L., and Stephenson, L.W. (1986). Canine latissimus dorsi hydraulic pouches: potential for left ventricular assistance. *Journal of Thoracic and Cardiovascular Surgery*, **91**, 534–44.

Mannion, J.D., Velchik, M., Hammond, R., Alavi, A., Mackler, T., Duckett, S. *et al.*

(1989). Effects of collateral blood vessel ligation and electrical conditioning on blood flow in dog latissimus dorsi muscle. *Journal of Surgical Research*, **47**, 332–40.

Mannion, J. D., Blood, V., Bailey, W., Bauer, T. L., Magno, M. G., DiMeo, F. *et al.* (1996). The effect of basic fibroblast growth factor on the blood flow and morphologic features of a latissimus dorsi cardiomyoplasty. *Journal of Thoracic and Cardiovascular Surgery*, **111**, 19–28.

Marber, M.S. (1994). Stress proteins in myocardial protection. *Clinical Science*, **86**, 375–81.

Mathes, S.J. and Nahai, F. (1981). Classification of the vascular anatomy of muscles: experimental and clinical correlation. *Plastic and Reconstructive Surgery*, **67**, 177–87.

Mayne, C.N., Jarvis, J.C., and Salmons, S. (1991). Dissociation between metabolite levels and force fatigue in the early stages of stimulation-induced transformation of mammalian skeletal muscle. *Basic and Applied Myology*, **1**, 63–70.

McMinn, R.M.H. and Vrbová, G. (1964). The effect of tenotomy on the structure of fast and slow muscle in the rabbit. *Quarterly Journal of Experimental Physiology*, **49**, 424–9.

Mocek, F.W., Anderson, D.R., Pochettino, A., Hammond, R.L., Spanta, A., Ruggiero, R. *et al.* (1992). Skeletal muscle ventricles in circulation long-term: 191 to 836 days. *Journal of Heart and Lung Transplantation*, **11**, S334–40.

Moreira, L.F.P., Bocchi, E.A., Stolf, N.A.G., Pileggi, F., and Jatene, A.D. (1993). Current expectations in dynamic cardiomyoplasty. *Annals of Thoracic Surgery*, **55**, 299–303.

Murry, C.E., Jennings, R.B., and Reimer, K.A. (1986). Pre-conditioning with ischaemia: a delay of lethal cell injury in ischaemic myocardium. *Circulation*, **74**, 1124–36.

Petrofsky, J.S. and Hendershot, D.M. (1984). The interrelationship between blood pressure, intramuscular pressure, and isometric endurance in fast and slow twitch skeletal muscle in the cat. *European Journal of Applied Physiology*, **53**, 106–11.

Polla, B.S., Mini, N., and Kahtengwa, S. (1991). Heat shock and oxidative injury in human cells. In *Heat shock* (ed. B. Maresca and D. Lindquist), pp. 279–90. Springer-Verlag, Berlin.

Radermecker, M.A., Sluse, F.E., Focant, B., Reznik, M., Fourny, J., and Limet, R. (1991). Influence of tension reduction, and peripheral dissection on histologic, biochemical and bioenergetic profiles, and kinetics of skeletal muscle fast-to-slow transformation. *Journal of Cardiac Surgery*, **6**, 195–203.

Radermecker, M.A., Triffaux, M., Fissette, J., and Limet, R. (1992). Anatomical rationale for use of the latissimus dorsi flap during the cardiomyoplasty operation. *Surgical and Radiological Anatomy*, **14**, 5–10.

Salmons, S. (1987). Biochemical and morphological evidence of increased protein and nucleic acid synthesis in chronically stimulated rabbit skeletal muscle. *Journal of Anatomy*, **152**, 229.

Salmons, S. and Jarvis, J.C. (1992). Cardiac assistance from skeletal muscle: a critical appraisal of the various approaches. *British Heart Journal*, **68**, 333–8.

Salmons, S. and Sréter, F.A. (1976). Significance of impulse activity in the transformation of skeletal muscle type. *Nature, London*, **263**, 30–4.

Salmons, S., Jarvis, J.C., Mayne, C.N., Chi, M.M.-Y., Manchester, J.K., McDougal, D.B., Jr *et al.* (1996). Changes in ATP, phosphocreatine and 16 metabolites in rabbit muscles stimulated for up to 96 hours. *American Journal of Physiology*, (in press).

Salo, D.C., Donovan, C.M., and Davies, K.J.A. (1991). HSP70 and other possible heat shock or oxidative stress proteins are induced in skeletal muscle, heart and liver during exercise. *Free Radical Biology and Medicine*, **11**, 239–46.

Scelsi, R. and Scelsi, L. (1995). Morphological analysis of skeletal muscle in patients with chronic heart failure. *Basic and Applied Myology*, **5**, 359–64.

Schreuder, J.J., Veen, F.H. van der, Velde, E.T. van der, Delahaye, F., Alfieri, O., Jegaden, O. *et al.* (1995). Beat-to-beat analysis of left ventricular pressure-volume relation and stroke volume by conductance catheter and aortic model-flow in cardiomyoplasty patients. *Circulation*, **91**, 2010–17.

Stuart, D.G. and Enoka, R.M. (1983). Motoneurons, motor units, and the size principle. In *The clinical neurosciences* (ed. R.N. Rosenberg), Section 5. *Neurobiology* (ed. W.D. Willis), pp. 471–517. Churchill Livingstone, New York.

Tabary, J.C., Tabary, C., Tardieu, C., Tardieu, G., and Goldspink, G. (1972). Physiological and structural changes in the cat's soleus muscle due to immobilization at different lengths by plaster casts. *Journal of Physiology*, **224**, 231–44.

Tatjanchenko, V.K. and Sherstennikov, E.N. (1978). Arterial architectonics of musculus latissimus dorsi in man and dog. *Archiv Anatomii Gistologii i Embriologii*, **74**, 28–33.

Taylor, G.I. and Palmer, J.H. (1987). The vascular territories (angiosomes) of the body: experimental study and clinical applications. *British Journal of Plastic Surgery*, **40**, 113–41.

Thomas, G.A., Isoda, S., Hammond, R.L., Lu, H., Nakajima, H., Nakajima, H.O. *et al.* (1996). Pericardium-lined skeletal muscle ventricles: two-year in-circulation experience. *Annals of Thoracic Surgery*, in press.

Tobin, G.R., Schusterman, B.A., Peterson, G.H., Nichols, G., and Bland, K.I. (1981). The intramuscular neurovascular anatomy of the latissimus dorsi muscle: the basis for splitting the flap. *Plastic and Reconstructive Surgery*, **67**, 637–41.

Vandenburgh, H.H., Hatfaludy, S., Karlisch, P., and Shansky, J. (1989). Skeletal muscle growth is stimulated by intermittent stretch-relaxation in tissue culture. *American Journal of Physiology*, **256**, C674–82.

Vrbová, G. (1963). Changes in the motor reflexes produced by tenotomy. *Journal of Physiology*, **166**, 241–50.

Weber, F.E. and Pette, D. (1990). Changes in free and bound forms and total amount of hexokinase isozyme II of rat muscle in response to contractile activity. *European Journal of Biochemistry*, **191**, 85–90.

Williams, N.S., Hallan, R.I., Koeze, D.H., and Watkins, E.S. (1989). Construction of a neorectum and a neoanal sphincter following previous proctolectomy. *British Journal of Surgery*, **76**, 1191–4.

Yellon, D.M., Alkhulaifa, A.M., and Pagsley, W.B. (1993). Preconditioning the ischaemic myocardium. *Lancet*, **342**, 276–7.

Index